Primary and Secondary Manufacturing of Polymer Matrix Composites

Primary and Secondary Manufacturing of Polymer Matrix Composites

Edited by
Kishore Debnath
Inderdeep Singh

CRC Press
Taylor & Francis Group
Boca Raton London New York

CRC Press is an imprint of the
Taylor & Francis Group, an **informa** business

CRC Press
Taylor & Francis Group
6000 Broken Sound Parkway NW, Suite 300
Boca Raton, FL 33487-2742

First issued in paperback 2019

© 2018 by Taylor & Francis Group, LLC
CRC Press is an imprint of Taylor & Francis Group, an Informa business

No claim to original U.S. Government works

ISBN-13: 978-1-4987-9930-0 (hbk)
ISBN-13: 978-0-367-88492-5 (pbk)

**Visit the Taylor & Francis Web site at
http://www.taylorandfrancis.com**

**and the CRC Press Web site at
http://www.crcpress.com**

Contents

Editors

Dr. Kishore Debnath is an assistant professor in the Department of Mechanical Engineering at the National Institute of Technology, Meghalaya, Shillong, India. He received his BE and MTech in mechanical engineering from the National Institute of Technology Agartala, Jirania, India, and the National Institute of Technology, Rourkela, India, respectively. Dr. Debnath earned his PhD from the Indian Institute of Technology Roorkee, India, in 2015. His research interests include composite materials, conventional and non-conventional machining and manufacturing processes. He has published 38 research papers in various national and international journals of repute and conference proceedings. He has also contributed 6 chapters in different books published by internationally renowned publishers.

Dr. Inderdeep Singh completed his bachelor's degree in mechanical engineering from the National Institute of Technology Hamirpur, India in 1998. He was first in the class of mechanical engineering. He completed his Master of Technology and PhD from the Indian Institute of Technology Delhi, India in 2000 and 2004, respectively. His doctoral research work was awarded by the Foundation for Innovation and Technology Transfer (FITT) as the 'Best Industry Relevant PhD Project' during the session 2004–2005. He started his career as a lecturer at the Institute of Technology (BHU) Varanasi, India in 2004. He joined the Indian Institute of Technology, Roorkee, India as a lecturer in 2005 and is currently working as an associate professor. He has published more than 180 research papers in journals and conferences. He has contributed 15 chapters in different books published by internationally renowned publishers. He has supervised 10 PhD candidates and is currently supervising 5 PhD scholars. He has guided 35 master's thesis and 32 BTech projects in the area of conceptualisation, processing and characterisation of composite materials. His current research focus is on developing fully biodegradable green composites.

Contributors

Jasti Anurag
Department of Mechanical Engineering
National Institute of Technology,
    Rourkela
Rourkela, India

Pramendra Kumar Bajpai
Division of Manufacturing Process and
    Automation Engineering
Netaji Subhas Institute of Technology
New Delhi, India

Sandhyarani Biswas
Department of Mechanical Engineering
National Institute of Technology,
    Rourkela
Rourkela, India

Vijay Chaudhary
Division of Manufacturing Process and
    Automation Engineering
Netaji Subhas Institute of Technology
New Delhi, India

Ranchan Chauhan
Faculty of Engineering and Technology
Shoolini University
Solan, India

Kishore Debnath
Department of Mechanical Engineering
National Institute of Technology,
    Meghalaya
Shillong, India

Siva Bhaskara Rao Devireddy
Department of Mechanical Engineering
National Institute of Technology,
    Rourkela
Rourkela, India

Anders E.W. Jarfors
Department of Materials and
    Manufacturing
School of Engineering
Jönköping University
Jönköping, Sweden

S. Kanagaraj
Department of Mechanical
    Engineering
Indian Institute of Technology
    Guwahati
Guwahati, India

Devarshi Kashyap
Department of Mechanical
    Engineering
Indian Institute of Technology
    Guwahati
Guwahati, India

U.K. Komal
Department of Mechanical and
    Industrial Engineering
Indian Institute of Technology Roorkee
Roorkee, India

J. Kumar
Department of Mechanical and
    Industrial Engineering
Indian Institute of Technology Roorkee
Roorkee, India

M.K. Lila
Department of Mechanical and
    Industrial Engineering
Indian Institute of Technology Roorkee
Roorkee, India

Sachin Maheshwari
Division of Manufacturing Process and
    Automation Engineering
Netaji Subhas Institute of Technology
New Delhi, India

Charan Mukundan
Department of Mechanical Engineering
Indian Institute of Technology
    Guwahati
Guwahati, India

K. Palanikumar
Department of Mechanical Engineering
Sri Sairam Institute of Technology
Chennai, India

Amar Patnaik
Department of Mechanical Engineering
Malaviya National Institute of
    Technology, Jaipur
Jaipur, India

T. Rajasekaran
School of Mechanical Engineering
SRM University
Chennai, India

M. Roy Choudhury
Department of Mechanical Engineering
National Institute of Technology,
    Meghalaya
Shillong, India

V.S. Senthil Kumar
College of Engineering
Anna University
Chennai, India

A.K. Sharma
Department of Mechanical and
    Industrial Engineering
Indian Institute of Technology Roorkee
Roorkee, India

Inderdeep Singh
Department of Mechanical and
    Industrial Engineering
Indian Institute of Technology Roorkee
Roorkee, India

Tej Singh
Department of Mechanical Engineering
Manav Bharti University
Solan, India

Jung Il Song
Department of Mechanical Engineering
Changwon National University
Gyeongsangnam-do, Korea

T.S. Srivatsan
Division of Materials Science and
    Engineering
Department of Mechanical Engineering
The University of Akron
Akron, Ohio

S. Suresh
Department of Mechanical Engineering
Velammal Engineering College
Chennai, India

S. Zafar
School of Engineering
Indian Institute of Technology Mandi
Kamand, India

1 Primary Manufacturing of Thermosetting Polymer Matrix Composites

Sandhyarani Biswas

CONTENTS

1.1 INTRODUCTION

Over the past few decades, it is found that polymers have replaced many of the conventional materials in various applications. Polymer matrix composite (PMC) mainly consists of a polymer resin as the matrix material, which is filled with reinforcement in the form of fibres or particles. This kind of material is used in the greatest diversity of composite applications due to its advantages such as lightweight, high stiffness, high specific strength, ease of processing and good corrosion resistance offered by the polymers over the conventional materials. In addition, the ability to manufacture products with the complex geometry using fewer components facilitates manufacturers to save the cost as compared to the same products made of conventional materials. The reinforcements (fibres and/or particles) are generally incorporated to enhance the desired properties of the composites. Although, the properties of the matrix and reinforcing phase have significant effect on the final properties of the composites, the processing technique selected to fabricate the composites also influences the performance of the composite materials. The objective of composite manufacturing is to fabricate the parts to obtain the best properties of the individual constituents or minimising their weaknesses. According to the end-item design requirements, a variety of manufacturing methods can be used for PMCs. Specific processes are generally used for primary manufacturing of PMCs, that is, near-net shape is processed thereby partially reducing the machining requirement of the fabricated components.

Generally, PMC fabrication processes utilise different types of raw materials, including resins, fibres, fabrics, mats, fillers and prepregs for manufacturing of composite products. Each technique needs different kinds of material systems, different tools and also different processing conditions. The most commonly used fabrication processes for PMCs include the hand lay-up, spray-up, vacuum bagging, compression moulding, injection moulding, resin transfer moulding (RTM), pultrusion and filament winding. Depending on the cost, property, quantity and quality of the product, a suitable technique can be selected. A brief outline of the different processing techniques mainly for thermosetting PMC is described in the following sections.

1.2   HAND LAY-UP PROCESS

Hand lay-up or wet lay-up process is an open moulding technique to fabricate polymer composite products wherein the resin and reinforcement are applied manually to prepare the laminated composite structure. It is the oldest, simplest and most commonly used method of fabricating PMCs. A release agent is generally applied onto the surface of the mould for preventing sticking of the part that leads to the easy removal of the finished composite products. A resin is mixed thoroughly with a curing agent and is applied on the surface of the mould. As per the requirement, the filler or other additives can also be used along with the resin. Then, the fibre is laid in, one layer at a time. Extra resin is generally squeezed out due to the applied pressure. While applying pressure, proper care should be taken as the non-uniform pressure may cause the development of composites with uneven thickness. Fibre displacement, resin rich area and void formations are some of the common defects that are mainly observed in the composites prepared through the hand lay-up process. Therefore, in order to squeeze out the excess resin, a roller is used that makes uniform distribution of the resin over the surface and removes air. Homogeneous fibre wetting is also obtained by the squeezing action of the roller. This process is repeated for all layers of reinforcement until the desired thickness is obtained. The reinforcement may be in the form of an aligned fabric such as woven rovings or chopped strand mat. The fibre should be uniformly positioned and also fit correctly on the shape of the mould surface to get the desired part profile. Then, the composite part is allowed for cure either at room temperature or at any specific temperature. The desired properties of the composites parts are obtained only after the complete curing. Finally, the fully hardened part is removed from the mould. The moulds can be made of plaster, wood, sheet metal and fibre-reinforced polymer (FRP) composites. Figure 1.1 shows the schematic of the hand lay-up process. Generally, the hand lay-up technique is suitable for any thermoset resin, which is available in the liquid form under the room temperature. Thermosetting resins such as vinyl ester, epoxy and unsaturated polyester are the commonly used polymers. Advantages of this method include its less

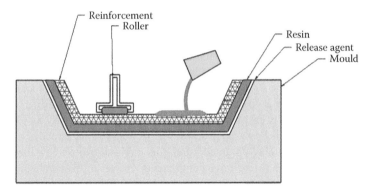

FIGURE 1.1 Hand lay-up process.

capital investment, easy to change mould/design and virtually no limit to the size of the part that can be produced. However, few limitations of this process include that the method is very time consuming, low-volume fraction/concentration of reinforcing phase, that is, maximum of 0.35 (Balasubramanian 2013), high content of voids and/or porosity (entrapped air bubbles), labour intensive (i.e. the quality of parts depends on the operator's skill and therefore inconsistent), only one finished surface which is in contact with the mould, control of thickness is not very accurate, to achieve the uniform fibre-to-resin ratio is difficult and also not suitable for high-volume production.

A large variety of products can be fabricated with this method, reaching from smallest parts to large coverings such as boat hulls, tanks and vessels, pick-up truck canopies, swimming pools, wind turbine blades and so on. In the 1940s, Americans compounded glass fibre (GF) and unsaturated polyester resin firstly and then produced military radar and aircraft fuel tank by hand lay-up process, which opened the path for the GF composite materials to be applied in military industry (Wang et al. 2011). Around a century ago, Ford had used this technique to manufacture automotive body using soya-based polymer-reinforced composites with natural fibre (Salit et al. 2015). Similarly, a great deal of work has been done by various researchers on the evaluation of various properties of PMCs manufactured by hand lay-up process. A conceptual design study has been performed on the development of a telephone stand produced with woven banana pseudo-stem fibre/epoxy composites (Sapuan and Maleque 2005). The same method has been used to fabricate a small multi-purpose table with woven banana pseudo-stem fibre/epoxy composites (Sapuan et al. 2007). The mechanical behaviour of a small boat with woven glass/sugar palm fibre-reinforced unsaturated polyester-hybrid composite fabricated using the hand lay-up technique has been studied (Misri et al. 2010). *Lantana camara* fibre-reinforced epoxy bio-composite pins are developed using hand lay-up technique for abrasive wear testing (Kalia 2016). The parameters effecting the quality and performance of fabricated composite products by the hand lay-up process include the fibre/resin type, fibre content, fibre orientation, pressure, curing time and so on. The influence of different parameters on the performance of fibre-reinforced PMCs made by this method has been analysed by many researchers (Biswas et al. 2011; Mishra and Biswas 2016).

1.3   SPRAY-UP PROCESS

The spray-up process is generally similar to the hand lay-up technique, with the difference being in the way of applying the fibre and resin onto the mould. Generally, a spray gun is used in this process to deposit the chopped fibre and resin on the mould surface to build the successive composite laminations. The composite has to be rolled to remove the entrapped air and to give a smooth surface finish. The deposited materials are generally cured under standard atmospheric conditions. Figure 1.2 shows the schematic of the spray-up process. This is a more cost-effective and much faster technique than hand lay-up. In addition, this method involves low tooling cost and is suitable for manufacturing large components. However, the mechanical properties are less because of the use of only randomly oriented short fibres. In addition, the resin needs to be low in viscosity and easily sprayable. Similar to the hand lay-up technique, this process also produces parts with good surface finish only on one side. The products that can be produced using this method includes the automobile and truck body parts, large structural panels, swimming pools, storage tanks, furniture components, bathtubs, boat hulls and so on.

The factors such as fibre type, fibre length, resin type, fibre volume ratio and so on are mainly influencing the performance of composites made by the spray-up technique. Fabrication of PMCs reinforced with both synthetic and natural fibres using this technique has been done by many researchers. The inter-laminar fracture toughness of carbon/epoxy composites by whisker (β-SiC)-reinforced interlamination fabricated by spray-up technique has been studied (Wang et al. 2002). Similarly, the mechanical behaviour of jute fibre-based polymer composites made by spray-up technique has been analysed (Kikuchi et al. 2014). In another study, the tensile properties of vinyl ester and unsaturated polyester-based composites reinforced with three types of fibre such as carbon fibre (CF), GF and basalt fibre are analysed (Yang and Wu 2016). In that study, 13 groups of specimens were prepared and tested to analyse their stress–strain curve, elastic modulus, tensile strength, breaking, elongation and other mechanical properties.

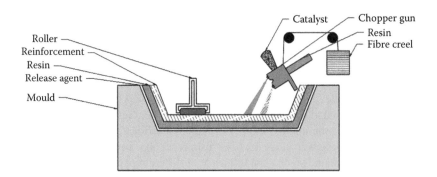

FIGURE 1.2   Spray-up process.

1.4 VACUUM-BAGGING PROCESS

The vacuum bagging is an open moulding process and also called as vacuum-bag moulding. In this process, the atmospheric pressure is used to hold the resin and fibres in place, which consolidates the layers within the laminate. The laminate is sealed in an airtight bag, and a vacuum pump then evacuates all the air out of the bag, resulting in an even atmospheric pressure over the entire laminate. The reinforcement can be either a woven mat or other fabric form. In vacuum-bagging process, layers from bottom include mould, release agent, composite laminate, peel ply, bleeder, release film, breather and vacuum bag. Sealant tapes are generally used on both sides of the bag to provide a vacuum-tight seal between the mould surface and the bag. Mould release agent is essential for preventing the resin from sticking to the mould surface when laminating a part. Peel ply is used to create a clean surface for bonding purpose. A bleeder layer is mainly used to absorb the excess resin from the laminate. The release film is a perforated film that allows the entrapped air and volatiles to escape. The purpose of the breather is to create uniform pressure around the part and also allows the air and volatiles to escape. The fabricated composites can be cured at room temperature or at an elevated temperature. Figure 1.3 shows the schematic of the vacuum-bagging process. The most commonly used polymer for vacuum-bagging process includes phenolic, epoxy and polyimide (PI). Parts such as aircraft structures, bathtub, large boat hulls, racing car components can be fabricated using vacuum-bagging process. Vacuum-bagging process has many advantages such as the higher fibre-to-resin ratio, better quality for the cost, very little emissions because the entire laminate is sealed in a bag, provides increased part consolidation and reduces mould costs. However, the disadvantages include that the breather clothe has to be replaced frequently and the process needs expensive curing ovens.

Fabrication of qualitative composite parts using vacuum-bag moulding process is dependent on many parameters such as viscosity of the resin, type of reinforcement, quality of sheets used, defined pressure and so on. The effect of various parameters on the performance of PMCs has been studied by a few researchers. The mechanical behaviour of bagasse fibre/polyester composites fabricated using vacuum-bagging process has been studied (Mariatti and Abdul Khalil 2009). The properties of the composite materials such as tensile, impact and flexural strength have been determined.

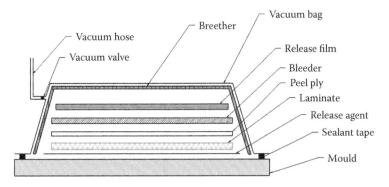

FIGURE 1.3 Vacuum-bagging process.

A study on the continuous GF-based polymer composites fabricated by vacuum bag moulding process has been done (Aparna et al. 2016). The study revealed that the vacuum bag moulding technique improves quality of the composites in terms of excellent surface finish, fibre-to-volume ratio, improved strength and stiffness and so on.

1.5   COMPRESSION-MOULDING PROCESS

Compression moulding is one of the most commonly used PMC processing techniques. It is a closed moulding process, where the composite product is fabricated inside the cavity of a die. In this process, the pre-determined amount of moulding compound or charge is generally placed in the lower half of a pre-heated mould cavity, which is then squeezed under heat and pressure for a specific period of time. During this period, the polymer resin is cured and the part is consolidated. The charge placed in between the two halves of a mould flows due to the application of heat and pressure and thus, obtain the shape of the mould cavity. After curing, the pressure is generally released and the mould is opened to remove the composite product. Curing of the composite can be done either at room temperature or at elevated temperature. The curing time mainly depends on the factors such as part thickness, resin type and mould temperature. For this process, the sheet moulding and bulk moulding compounds are the most commonly used materials. For making the moulding compounds, the raw materials used are the resin, curing agent, fibre, filler, release agent and other additives as per the requirement. The schematic of the compression-moulding process is shown in Figure 1.4. The process is suitable for manufacturing both thermosetting and thermoplastic polymer-based products. Unsaturated polyester, vinyl ester, epoxy, polyvinyl ester and phenolic resins are the most commonly used polymers. While, among the thermoplastic polymers, polycarbonate (PC), polyvinyl chloride (PVC), polypropylene (PP), polyetheretherketone (PEEK) can be used. Typical products that can be produced using this process include the electrical fixtures, control boxes, machine guards, bumper beams, road wheels, refrigerator doors, furniture, door panels, hoods, kitchen bowls and trays, automobile panels and so on. The main advantages of compression moulding include that the process needs short cycle time, that is, faster production, volume of production is high, both interior and exterior surfaces are finished, better quality surface, uniformity in part shape, less maintenance

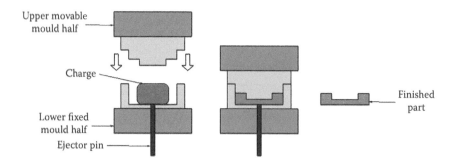

FIGURE 1.4   Compression-moulding process.

cost, better dimensional accuracy, very little finishing operation is required and better control of fibre content. However, the high initial capital investment, not economical for low volume of production, not suitable for large-sized parts, and limitation on mould depth are the major disadvantages of this method.

The parameters that are mainly affecting the performance of composites made by the compression moulding are the quantity of moulding compound, pressure of the moulding process, mould temperature, type of resin used, cure time and so on. A great deal of work has already been done on the influence of various factors on the properties of PMCs by many researchers. The structure and properties of composites made with polyurethane (PU) pre-polymer and various soy products using compression moulding has been analysed (Chen et al. 2003). A series of sheets from soy products such as soy dreg, soy protein isolate and soy whole flour, respectively, has been fabricated with PU pre-polymer. The effect of fibre orientation on mechanical properties like tensile and flexural strength of compression moulded sisal fibre reinforced epoxy composites has been studied (Kumaresan et al. 2015). Similarly, a comparative study has been made on the compression moulded-unsaturated polyester composites reinforced with GF and flax fibres, respectively (Holbery and Houston 2006). The study revealed that at the same fibre loading, the properties are comparable. A review has been made on the compression moulding of natural fibre reinforced thermoset composites in terms of their thermal and mechanical properties (Ismail et al. 2015). Similarly, the mechanical properties of scaffolds using chitosan–polyester blends and composites fabricated using compression moulding have been studied (Correlo et al. 2009). The mechanical properties of red mud filled sisal and banana fibre reinforced polyester composites fabricated by compression moulding has been studied (Prabu et al. 2012).

1.6 INJECTION-MOULDING PROCESS

Injection moulding is one of the widely used PMC processing techniques. It is a high-volume production process and is suitable to produce short fibre-reinforced PMCs. In this process, the moulding compound such as polymer pellets, short fibres or fillers are fed into the barrel through the feed hopper in which the polymer softens. The complete mixing and uniform distribution of fibres into the resin are ensured through the rotation of screw. The heating process in heated barrel and shearing action of a reciprocating screw causes the moulding compound to melt. The blend is injected into the mould cavity with pressure for a specific period of time through the nozzle, which is attached to the end of the barrel. As per the requirement, the mould may contain single or multiple cavities. After cooling, the composite product is taken out of the mould cavity. Generally, the injection-moulding process produces composite products almost always to net shape. Figure 1.5 shows the schematic of the injection-moulding process. There are two basic types of plastic injection moulding such as thermoplastic and thermoset. In the former, a thermoplastic polymer is generally melted and forced into the mould through an orifice, which is kept relatively cool. Then, the material solidifies in the mould from which it can finally be removed. However, in the case of thermoset injection moulding, the solidification takes place at high temperature. Injection-moulding technique can be used to produce parts such as vanes, engine cowling defrosters and aircraft radomes, auto parts and so on. The advantages of injection-moulding process

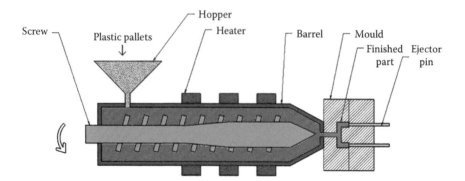

FIGURE 1.5   Injection-moulding process.

are that it is a relatively faster production process, that is, suitable for high production, very little post-production work required, full automation is also possible, provides good dimensional control, produces complex shapes and so on. However, the high initial capital investment is the major limitation of this process.

The performance and quality of the injection moulded composite materials are generally influenced by several parameters such as injection pressure, temperature in the barrel, cross-sectional area of nozzle, mould temperature, screw speed, injection time, holding time and so on. Injection mould performance of machined ceramic filled epoxy tooling boards has been studied (Tomori et al. 2004). They have evaluated the effects of the composite tooling board material formulation and machining parameters on mould performance and the quality of injection molded parts. Mould performance has also been evaluated based on the mould condition during molding, the molded part flexural strength and surface roughness, and the consistency of these parameters over the duration of the study. The effect of mould temperature and process time on the degree of cure of glass fibre reinforced epoxy based compounds for thermoset injection moulding and prepreg compression moulding has been studied (Deringer et al. 2017). An experimental study of the injection moulding of an unsaturated polyester bulk-moulding compound including rheology, mould filling, and fibre orientation has been done (Paiva et al. 2015). A numerical simulation of the filling stage, taking into account non-Newtonian behaviour and cure reactions has been described. A comparison between the numerical and experimental results has also been done. The development of an epoxy resin system for the injection moulding of long-fibre epoxy composites has been made (Ishida and Zimmerman 1994). The impact strength and solvent resistance of the long-fibre composite has been examined. The properties of the thermoset long-fibre epoxy composite has also been compared to those of a thermoplastic injection molded long-fibre phenoxy composite.

1.7   RESIN TRANSFER-MOULDING PROCESS

RTM is a novel technique for manufacturing of high-performance thermoset polymer composites. It is a closed moulding process. In this process, the fibre is generally packed in the cavity of a closed mould as per the required geometrical arrangement.

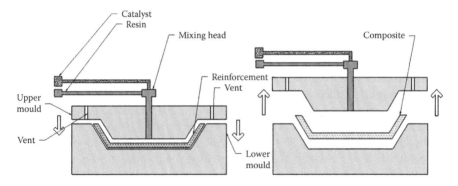

FIGURE 1.6 Resin transfer-moulding process.

Then a liquid resin with catalyst is injected into it at high pressure and temperature (curing). The pressure of injection wets the fibres. Vents are generally provided to release the gases in the mould cavity. The mould is required to properly close and clamp. The part is then cured either at room temperature or under heat and pressure for several hours. The cure may be initiated either by heating the mould or by adding the inhibitors to the resin. Then the mould is opened and part is removed once it is sufficiently hardened. Cure cycle is dependent on many parameters such as the type of resin, part thickness and the temperature of the mould and so on. Figure 1.6 shows the schematic of RTM process. The raw materials generally used for fabrication of PMCs by this process are resins, fibres, accelerators, catalysts and additives. The most commonly used resins in this technique are epoxy, vinyl ester, polyester and phenolic. Reinforcement in the form of either woven mat or strand mat can be used. This process is suitable for manufacturing small- to large-sized parts in small to medium quantities. The process is commonly used in aerospace, automotive and sporting industry. The main advantages of RTM include that the production cycle is much faster than the hand lay-up process, good surface finish is possible on both sides, better control over the product thickness, short cycle time, volatile emissions are low, better utilisation of resin and fibre, which reduces material wastage. However, few limitations of the process include that the mould cavity generally limits the size of the part, tooling cost is high, limited to low viscosity resins and post-trimming is required.

The performance of final composite product is influenced by many parameters such as resin viscosity, fibre geometry, fibre content, mould temperature, applied pressure and so on. The viscosity of the resin has a significant influence on the RTM process because the injection time mainly depends on the viscosity of the polymer resin. If the viscosity of resin is high, then the high pressure is necessary that may cause displacement of fibres. The effect of various factors on the properties of PMCs has been evaluated by many researchers. A modelling of RTM of FRP composites with oriented unidirectional plies has been carried out (Cairns et al. 1999). The thermophysical behaviour of hemp/kenaf fibre-reinforced unsaturated polyester composites fabricated by RTM process has been studied (Rouison et al. 2004). The curing behaviour of the composite materials has also been predicted using a curing model. The study revealed that the temperature in the centre of the mould has been increased significantly compared to other composite materials due to the lower thermal conductivity value. In another study, the mechanical behaviour of sisal fibre/polyester textile composites fabricated by RTM process has been analysed (Pothan et al. 2008). The effect of weave architecture, resin viscosity, applied pressure and surface modification of fibre on the properties of composites has been studied. The influence of fibre treatment on the mechanical and water absorption properties of sisal fibre/polyester composites made by RTM process has been studied (Sreekumar et al. 2009). The study revealed that the impact properties of the composites decrease as compared to the untreated ones. It has also been reported that the treatment decreases the water uptake of the composite materials, which supports the greater fibre/matrix interaction. The impact of stitching density of non-woven fibre mat on the mechanical behaviour of kenaf/epoxy composites made by this method has been investigated (Salim et al. 2011).

1.8 FILAMENT-WINDING PROCESS

Filament winding is a PMC fabrication technique mainly used to produce products of axisymmetric shapes. It is a continuous process in which the fibres are wound around a rotating mandrel in specific orientations. The fibre spools are generally mounted to a creel and fed through an alignment device in order to create a band of fibres. Generally, the fibres are pulled through a resin bath immediately before being wound onto the rotating mandrel. This process is repeated to form additional layers until the desired part thickness is obtained. The fibre orientation and thickness can therefore be varied to obtain the optimum performance. The desired fibre angle can be obtained by controlling the motion of the mandrel and carriage unit. The tension applied to the reinforcement over the mandrel generally creates a positive pressure, which compact the laminates. This tension can also be controlled to achieve optimum fibre content and good consolidation of the laminate. Parts are cured either at room temperature or at elevated temperatures. The mandrel is generally extracted from the composite part after curing and then reused. Figure 1.7 shows the schematic of the filament-winding process. Different kinds of winding patterns can be used in filament-winding technique. Figure 1.8 shows three most commonly

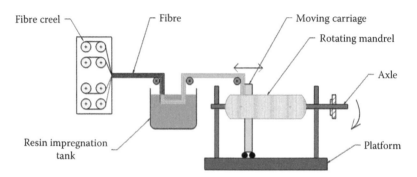

FIGURE 1.7 Filament-winding process.

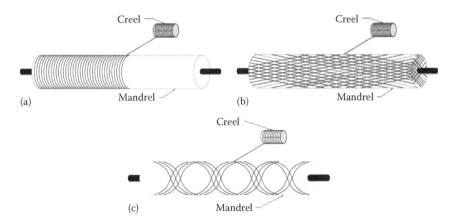

FIGURE 1.8 Filament-winding patterns: (a) hoop winding, (b) polar winding and (c) helical winding.

used winding patterns. Similarly, different materials such as aluminium alloys or steel with suitable dimensions are appropriate for making the mandrels. However, depending on the part geometry and complexity, other types of mandrel are also used such as soluble plaster, collapsible rubber and so on. Thermosetting resins such as epoxy, polyester and phenolic are the most commonly used resins for filament-winding technique. This process is normally used to produce the parts such as pipes, pressure vessels, wind turbine blades, shafts, storage tanks and other aerospace components. The process is very fast and highly automated. The structural properties of laminates can be very good because the straight fibres can be laid in a complex pattern in order to match the applied loads. High production rate, excellent mechanical properties, good thickness control, better control of fibre orientation and content, high volume fraction/concentration of reinforcing phase and good internal finish are also the other advantages of this process. However, the process is suitable for limited range of shapes specifically to convex-shaped components. The external surface of the fabricated part is cosmetically unattractive; high mandrel costs for large components and the need of low viscosity resins are also the disadvantages of this process.

In filament-winding technique, the parameters that are affecting the quality of the part are fibre volume fraction, size range, processing temperature, fibre tension, winding speed, design of resin bath, viscosity of resin, mandrel type/material, winding pattern, carriage movement and so on. The impact of various parameters on the performance of PMCs fabricated by this technique has been studied by many researchers. The effect of filament-winding parameters on the composite vessel strength and quality has been evaluated (Cohen 1997). The stress and deformation of multi-layered filament-wound CF/epoxy composite pipes have been analysed (Xia et al. 2001). Detailed stress and strain distributions for the three different angle-ply pipe designs under the internal pressure load are evaluated. The impact of filament winding tension on the physical and mechanical behaviour of GF-reinforced polymeric-composite tubular has been studied (Mertiny and Ellyin 2002). An experimental investigation on the influence of multi-angle filament winding on the strength properties of tubular composite structures has been carried out (Mertiny et al. 2004). In another investigation, kenaf fibre/unsaturated polyester composite hollow shafts were fabricated using filament-winding technique and their mechanical behaviour has been studied (Misri et al. 2015). A continuous kenaf fibre rovings were pulled through a drum-type resin bath and wound around an aluminium-rotating mandrel. The experimental results are compared with the results obtained from finite element analysis.

1.9   PULTRUSION PROCESS

Pultrusion is one of the most cost-effective processes for fabricating structural composite profile. It is a continuous process, producing product profile of uniform cross section. This process is similar to extrusion; however, the main difference being that the composite part is pulled from the die, whereas in case of extrusion process it is pushed through the die. The process was developed by W. Brant Goldsworthy in the late 1950s and was patented as a process to make a fishing rod (Fairuz et al. 2014b). In this process, the fibres are generally pulled through a resin bath and then passed through a heated die to form the desired composite profiles. The die completes the impregnation of the fibres, controls the content of resin and also cures the composite products into the final shape. The polymerisation takes place inside the heated dies. The shape of the final composite product generally follows the shape of the dies. The most common shapes can be obtained are square, rectangular, circular and I-shaped sections. The end products are usually in the form of bars and rods. Figure 1.9 shows the schematic of the pultrusion process. This is a very fast and highly automated process. Other advantages include that the high volume fraction is possible, not labour intensive, better strength properties, consistent quality possible, good structural properties, better surface finish, resin content can be controlled easily and so on. However, the main disadvantages include that the process is suitable to produce components with uniform cross sections, high cost of die and more suitable for thermosetting polymers. The most commonly used thermosetting polymers for this process include unsaturated polyester and epoxy. Pultrusion can be used to produce parts such as tubing, solid rods, long flat sheets, tool handles, different structural sections and so on.

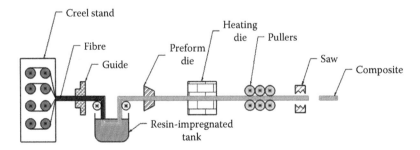

FIGURE 1.9   Pultrusion process.

The properties of the PMCs fabricated by pultrusion are influenced by the various processing parameters such as fibre fraction, resin viscosity, resin polymerisation, die temperature and pulling speed. Therefore, in order to produce a high-quality product, it is important to control these process parameters. The influence of various parameters on the performance of composites has already been done by previous researchers. A comprehensive analytical and experimental program for the analysis and design of the pultruded FRP composite beams under bending has been studied (Davalos et al. 1996). The study revealed that the material architecture of pultruded composite shapes can be efficiently modelled as a layered system. An iterative approach has been developed using both heat-transfer and curing sub-models for simulation of the pultrusion for PMCs (Suratno et al. 1998). In order to simulate the heat transfer and temperature profiles inside the die, a two-dimensional finite element model has been applied. The pultrusion process of carbon/epoxy composite rods has been simulated and the results are compared with the results obtained in their previous work. A theoretical modelling of pulling speed for pultruded epoxy composites has been carried out (Moschiar et al. 1996). The study revealed that high pulling velocity reduces the thermal stress and pressure inside the heated die. The mechanical behaviour of GF-based polymer composites fabricated by pultrusion using chemically modified soy-based epoxy resins has been studied (Zhu et al. 2004). The investigation revealed that the pultruded composites with soy-based co-resin systems show comparable or improved structural performance characteristics such as impact resistance, flexural strength and modulus. The mechanical behaviour of kenaf fibre/vinyl ester composites fabricated by pultrusion process using Taguchi's design of experiment has been studied (Fairuz et al. 2014a). The effect of three different parameters such as speed, temperature and filler loading has been analysed. Although, the pultrusion process is suitable for producing products using thermosetting polymers; however, studies on the thermoplastic composites have also been done by various researches. The possibility of the use of flax fibre as reinforcement in thermoplastic-based pultruded composites has been studied (Van de Velde and Kiekens 2001). Round flax/PP and rectangular glass/PP profiles are fabricated using pultrusion process and tested for various properties.

1.10 CONCLUSION

Over the years, there have been many developments in the PMCs fabrication processes. Some of the recent improvements include more advanced and sophisticated equipment that has been used for vacuum-bag moulding technology with precise control of pressure, temperature and other parameters to achieve better quality products. Similarly, many variations in compression moulding have been developed that are suitable for many engineering applications. A number of variations of the conventional injection-moulding process have been developed. Gas-assisted injection moulding and reaction-injection moulding are the examples of improved version of the process. Many improvements have been made in RTM process. An improved version of this process is vacuum-assisted RTM. Utilisation of vacuum in this method generally helps in removing the entrapped air and also the better flow of polymer resin through the fibre packing. Nowadays, the filament-winding equipments are available with numerous levels of sophistication. The machine generally ranges from the simple mechanically controlled equipment to sophisticated computer-controlled equipment with three or four axes of motion. Though, filament-winding process was initially developed for only thermoset resins, however winding with thermoplastic resins has also been practiced. The significant technological developments have greatly enhanced the quality of moulded parts and the efficiency of the PMC manufacturing processes.

REFERENCES

Arumuga Prabu, V., V. Manikandan, M. Uthayakumar, and S. Kalirasu. 2012. Investigations on the mechanical properties of red mud filled sisal and banana fiber reinforced polyester composites. *Materials Physics and Mechanics* 15: 173–179.

Aparna, M.L., G. Chaitanya, K. Srinivas, and J.A. Rao. 2016. Fabrication of continuous GFRP composites using vacuum bag moulding process. *International Journal of Advanced Science and Technology* 87: 37–46.

Creighton, C.J., and T.W. Clyne. 2000. The compressive strength of highly-aligned carbon-fibre/epoxy composites produced by pultrusion. *Composites Science and Technology* 60 (4): 525–533.

Balasubramanian, M. 2013. *Composite Materials and Processing*. Boca Raton, FL: CRC Press.

Biswas, S., B. Deo, A. Patnaik, and A. Satapathy. 2011. Effect of fiber loading and orientation on mechanical and erosion wear behaviors of glass-epoxy composites. *Polymer Composites* 32(4): 665–674.

Cairns, D.S., D.R. Humbert, and J.F. Mandell. 1999. Modeling of resin transfer molding of composite materials with oriented unidirectional plies. *Composites Part A: Applied Science and Manufacturing* 30(3): 375–383.

Chen, Y., L. Zhang, and L. Du. 2003. Structure and properties of composites compression-molded from polyurethane prepolymer and various soy products. *Industrial and Engineering Chemistry Research* 42(26): 6786–6794.

Cohen, D. 1997. Influence of filament winding parameters on composite vessel quality and strength. *Composites Part A: Applied Science and Manufacturing* 28(12): 1035–1047.

Correlo, V.M., L.F. Boesel, E. Pinho, A.R. Costa-Pinto, M.L. Alves da Silva, M. Bhattacharya, J.F. Mano, N.M. Neves, and R.L. Reis. 2009. Melt-based compression-molded scaffolds from chitosan-polyester blends and composites: Morphology and mechanical properties. *Journal of Biomedical Materials Research Part A* 91A(2): 489–504.

Davalos, J.F., H.A. Salim, P. Qiao, R. Lopez-Anido, and E.J. Barbero. 1996. Analysis and design of pultruded FRP shapes under bending. *Composites Part B: Engineering* 27(3): 295–305.

Deringer, T., C. Gröschel, and D. Drummer. 2017. Influence of mold temperature and process time on the degree of cure of epoxy-based materials for thermoset injection molding and prepreg compression molding. *Journal of Polymer Engineering*: 1–9.

Ishida, H., and D.A. Zimmerman. 1994. The development of an epoxy resin system for the injection molding of long-fiber epoxy composites. *Polymer Composites* 15(2): 93–100.

Fairuz, A.M., S.M. Sapuan, E.S. Zainudin, and C.N. Aiza. 2014a. Study of pultrusion process parameters. In *Proceedings of the Postgraduate Symposium on Composites Science and Technology 2014 & 4th Postgraduate Seminar on Natural Fibre Composites*, Putrajaya, Malaysia, January 28, pp. 116–120.

Fairuz, A.M., S.M. Sapuan, E.S. Zainudin, and C.N.A. Jaafar. 2014b. Polymer composite manufacturing using a pultrusion process: A review. *American Journal of Applied Sciences* 11(10): 1798–1810.

Holbery, J., and D. Houston. 2006. Natural-fiber-reinforced polymer composites in automotive applications. *JOM* 58(11): 80–86.

Ismail, N.F., A.B. Sulong, N. Muhamad, D. Tholibon, M.K.F. MdRadzi, and W.A.S. WanIbrahim. 2015. Review of the compression moulding of natural fiber-reinforced thermoset composites: Material processing and characterisations. *Journal of Tropical Agricultural Science* 38(4): 533–547.

Kalia, S., Ed. 2016. *Biodegradable Green Composites*. Hoboken, NJ: John Wiley & Sons.

Kumaresan, M., S. Sathish, and N. Karthi. 2015. Effect of fiber orientation on mechanical properties of sisal fiber reinforced epoxy composites. *Journal of Applied Science and Engineering* 18(3): 289–294.

Kikuchi, T., Y. Tani, Y. Takai, A. Goto, and H. Hamada. 2014. Mechanical properties of jute composite by spray up fabrication method. *Energy Procedia* 56: 289–297.

Mariatti, J., and H.P.S. Abdul Khalil. 2009. Properties of bagasse fibre-reinforced unsaturated polyester (USP) composites. In Sapuan, S.M., Ed., *Research on Natural Fibre Reinforced Polymer Composites*, pp. 63–83. Serdang, Malaysia: UPM Press.

Mertiny, P., and F. Ellyin. 2002. Influence of the filament winding tension on physical and mechanical properties of reinforced composites. *Composites Part A: Applied Science and Manufacturing* 33(12): 1615–1622.

Mertiny, P., F. Ellyin, and A. Hothan. 2004. An experimental investigation on the effect of multi-angle filament winding on the strength of tubular composite structures. *Composites Science and Technology* 64(1): 1–9.

Mishra, V., and S. Biswas. 2016. Three-body abrasive wear behavior of short jute fiber reinforced epoxy composites. *Polymer Composites* 37(1): 270–278.

Misri, S., Z. Leman, S.M. Sapuan, and M.R. Ishak. 2010. Mechanical properties and fabrication of small boat using woven glass/sugar palm fibres reinforced unsaturated polyester hybrid composite. *IOP Conference Series: Materials Science and Engineering* 11(1): 1–13.

Misri, S., S.M. Sapuan, Z. Leman, and M.R. Ishak. 2015. Torsional behaviour of filament wound kenaf yarn fibre reinforced unsaturated polyester composite hollow shafts. *Materials & Design* 65: 953–960.

Moschiar, S.M., M.M. Reboredo, H. Larrondo, and A. Vazquez. 1996. Pultrusion of epoxy matrix composites: Pulling force model and thermal stress analysis. *Polymer Composites* 17: 850–858.

Paiva, L.O.de., L. de Oliveira Couto, A.C.A. da Costa, M.C.A.M. Leite. 2015. Degradation of polyester-based and vegetable fiber polymeric composites. *International Journal of Scientific Research in Science and Technology* 4(1): 98–101

Pothan, L.A., Y.W. Mai, S. Thomas, and R.K.Y. Li. 2008. Tensile and flexural behavior of sisal fabric/polyester textile composites prepared by resin transfer molding technique. *Journal of Reinforced Plastics and Composites* 27(16–17): 1847–1866.

Rouison, D., M. Sain, and M. Couturier. 2004. Resin transfer molding of natural fiber reinforced composites: Cure simulation. *Composites Science and Technology* 64(5): 629–644.

Salim, M.S., Z.A.M. Ishak, and S. Abdul Hamid. 2011. Effect of stitching density of nonwoven fiber mat towards mechanical properties of kenaf reinforced epoxy composites produced by resin transfer moulding (RTM). *Key Engineering Materials* 471–472: 987–992.

Salit, M.S., M. Jawaid, N. Bin Yusoff, and M.E. Hoque. 2015. *Manufacturing of Natural Fibre Reinforced Polymer Composites*. Cham: Springer International Publishing.

Sapuan, S.M., N. Harun, and K.A. Abbas. 2007. Design and fabrication of a multipurpose table using a composite of epoxy and banrna pseudostem fibres. *Journal of Tropical Agriculture* 45(1–2): 66–68.

Sapuan, S.M., and M.A. Maleque. 2005. Design and fabrication of natural woven fabric reinforced epoxy composite for household telephone stand. *Materials & Design* 26(1): 65–71.

Sreekumar, P.A., S.P. Thomas, J. Marc Saiter, K. Joseph, G. Unnikrishnan, and S. Thomas. 2009. Effect of fiber surface modification on the mechanical and water absorption characteristics of sisal/polyester composites fabricated by resin transfer molding. *Composites Part A: Applied Science and Manufacturing* 40(11): 1777–1784.

Suratno, B.R., L. Ye, and Y.-W. Mai. 1998. Simulation of temperature and curing profiles in pultruded composite rods. *Composites Science and Technology* 58(2): 191–197.

Tomori, T., S. Melkote, and M. Kotnis. 2004. Injection mold performance of machined ceramic filled epoxy tooling boards. *Journal of Materials Processing Technology* 145: 126–133.

Van de Velde, K., and P. Kiekens. 2001. Thermoplastic pultrusion of natural fibre reinforced composites. *Composite Structures* 54(2): 355–360.

Wang, R.M., S.R. Zheng, and Y.G. Zheng. 2011. *Polymer Matrix Composites and Technology*. Cambridge: Woodhead Publishing.

Wang, W.X., Y. Takao, T. Matsubara, and H.S. Kim. 2002. Improvement of the interlaminar fracture toughness of composite laminates by whisker reinforced interlamination. *Composites Science and Technology* 62(6): 767–774.

Xia, M., H. Takayanagi, and K. Kemmochi. 2001. Analysis of multi-layered filament-wound composite pipes under internal pressure. *Composite Structures* 53(4): 483–491.

Yang, Z., and K. Wu. 2016. Experimental analysis of tensile mechanical properties of sprayed FRP. *Advances in Materials Science and Engineering* 2016: 1–12.

Zhu, J., K. Chandrashekhara, V. Flanigan, and S. Kapila. 2004. Manufacturing and mechanical properties of soy-based composites using pultrusion. *Composites Part A: Applied Science and Manufacturing* 35(1): 95–101.

2 Primary Manufacturing of Thermoplastic Polymer Matrix Composites

Kishore Debnath, M. Roy Choudhury and Anders E.W. Jarfors

CONTENTS

2.1   AN OVERVIEW

The fibre-reinforced thermoplastic composites have numerous advantages over thermoset composites due to their exclusive properties such as higher fracture toughness, unlimited self-life and easily recyclable. The manufacturing processes that are available for thermoset composites can be selectively adapted for thermoplastic composites. However, the primary manufacturing of thermoplastic composites is not very conducive as processing of thermoplastic composites requires heavy and strong tooling. In addition, sophisticated equipment is required to apply heat and pressure during processing of thermoplastic composites. A few cost-effective techniques have been developed for manufacturing of thermoplastic composites in order to obtain high-quality and high-definition products. In this chapter, the primary manufacturing processes (tape winding, pultrusion, compression moulding, injection moulding, autoclave processing, diaphragm forming and hot pressing) for thermoplastic composites have been comprehensively discussed. The basic processing steps, methods of applying heat and pressure, suitable raw materials, advantages and disadvantages of the processes have also been adequately discussed in context of thermoplastic composites.

2.2   INTRODUCTION

Composite materials are multi-phase material system, which consists of matrix and reinforcing materials. When the matrix material used in the composite is polymer then the composite is called polymer matrix composites (PMCs). The PMCs can be bifurcated into thermosetting and thermoplastic composites on the basis of the chemical reaction they undergo during the processing. Thermoset polymers undergo irreversible chemical reactions by creating cross-links between the molecular chains during curing. But thermoplastics do not undergo any chemical reaction and hence, no cross-links are formed between the molecules. The weak bonds between the molecules break when heat is applied. Subsequently, the movement of the molecules takes place relative to each other and hence, the molecules flow to a new configuration of a predefined shape on application of pressure. The solidification of the molecules takes place by restoring the intermolecular forces on cooling. Thermoplastics offer many advantages over thermoset polymers. Low processing cost, design flexibility and ease of moulding complex parts are some of the salient characteristics of the thermoplastic polymers. Moreover, thermoplastic polymers are essentially stronger, can be easily reprocessed and re-shaped as compared to the thermoset polymers.

Thermoplastic composites have immersed as a popular raw material in aerospace and automotive industry due to their high toughness, high production rate and less environmental effect characteristics. However, the manufacturing of thermoplastic composite is difficult as processing temperature for thermoplastic is high as compared to the thermoset. High melt viscosity and the lack of drape and tack characteristics of the thermoplastic prepregs also pose a challenge for their processing. The tooling required for manufacturing of thermoplastic composites is also heavy and expensive. The primary manufacturing of thermoplastic composite involves (1) wetting or impregnation, (2) lay-up, (3) consolidation and (4) solidification. The impregnation ensures the flow of resin around the fibres and thus, mixing of fibres and resin takes place. The major factors affecting the impregnation of fibre with the matrix are (1) viscosity, (2) surface tension and (3) capillary action. The viscosity of the thermoplastic polymer is quite high (104–108 cp) and hence, the processing of thermoplastic polymers requires relatively high pressure for better impregnation. In lay-up, the layers of the fibres and resin (or prepregs) are placed at desired angle to obtain the required thickness of the composite. The consolidation ensures the intimate contact at the interference between the prepreg layers that can be achieved by removing the entrapped air by applying adequate pressure. A poor consolidation part produces voids and dry spots, which results in poor quality of the composite part. The final stage is the solidification of composite parts. The solidification time required for curing of thermoplastics composite is less than a minute as no chemical changes occur during the process. There are several manufacturing techniques established for thermoplastic composites namely tape winding, pultrusion, compression moulding, injection moulding, autoclave processing, diaphragm forming and hot pressing. Though the four basic steps (impregnation, lay-up, consolidation and solidification) are common for all the manufacturing processes, they may vary in terms of applying heat and pressure.

2.3   TAPE WINDING

Thermoplastic tape winding is used to manufacture axisymmetric as well as non-axisymmetric components. Open (cylinders) or closed-end structures (pressure vessels or tanks) can be manufactured using this technique. Generally, continuous fibres are used as reinforcement for making the parts. Thermoplastic tape winding is also commonly known as thermoplastic filament winding. The raw materials used for processing through tape winding are given in Table 2.1. In this process, continuous fibres are laid down on the surface of a rotating mandrel along a predetermined path after soaking them in thermoplastic resin bath as shown in Figure 2.1. The heat and pressure are applied at the contact point of the roller and the rotating mandrel for melting and consolidation of thermoplastic. The pressure is applied using a consolidation roller. The various sources of heat supplied during consolidation are listed in Table 2.2. The roller pressure should be maintained until the temperature of the resin at the interface of the bonded surface drops below the melting temperature. The thermoplastic tape-winding process is also called online consolidation process because the incoming tape consolidates at the point where it is laid down. Hence, this process does not require any additional autoclave or convection oven for curing of the produced part.

TABLE 2.1

Raw Materials Suitable for Processing in Tap-Winding Process

Polymers	Polyetheretherketone (PEEK), polyphenylene sulfide (PPS), polyamide (nylon 6), polyetherimide (PEI), polypropylene (PP), polymethylmethacrylate (PMMA) and so on
Reinforcements	Carbon fibre, glass fibre, aramid fibre and so on

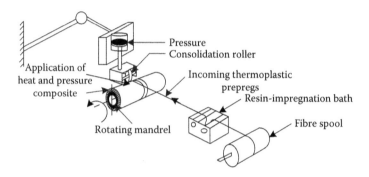

FIGURE 2.1   Schematic of thermoplastic tape winding.

TABLE 2.2

Various Sources of Heat Supplied During the Consolidation Process in Tape Winding

Heating Source	Method	Remarks
Hot rollers (induction or resistance heating) [1]	Provided from the top of the incoming tape	Sticking of material on the surface of the roller causes problems during rolling
Resistance heating [1]	By passing electric current through the carbon fibre	Not applicable for glass and aramid fibre
High-frequency waves [1]	By oscillating the molecules of the thermoplastic	Applicable for thermoplastic containing polar molecules
Open flame or acetylene gas torch [1]	Applied at consolidation point	Polymer degradation occurs
Hot air or hot nitrogen gas [2]		Cost effective but poor heat efficiency
Laser heating [3–5]		• Provide cleaner atmosphere but expensive • Localised heat at small area

Thus, the tooling cost and the total cycle time are significantly reduced, which makes the process more efficient. The thermoplastic tape-winding process mainly depends on the winding speed, heat intensity and consolidation pressure. The consolidation rate of a moulded part is characterised by the time required to established the intimate contact and hence the adhesive bonding between the reinforced fibre and the thermoplastic resin. The consolidation is difficult at higher speed due to the short dwell time. However, the degradation of the matrix occurs at low lay-down speed. The consolidation process is completed as the cooling and solidification of the product take place. The degree of crystallinity is highly affected by the cooling rate. As the cooling rate decreases, the tensile strength, compressive strength and solvent resistance of the matrix increase. Crystallinity is also affected by remelting, re-solidification, annealing and the speed of the consolidation roller. At a lower speed of consolidation roller, the degree of crystallinity is more. In the filament-winding process, the composite constituents experience repeated heating and cooling as additional layers are wound onto the rotating mandrel. In the hot nitrogen gas-aided tap-winding process, both the heating rate and cooling rate are higher than 20,000°C/min and 4,000°C/min, respectively, and shorter melting time is in the range of 1–4 s.

2.3.1   APPLICATIONS

Thermoplastic tape winding is not a commercially popular process. In recent years, many research works have been carried out to investigate the feasibility of tape-winding process for making prototype parts [6]. This process is basically suitable for manufacturing parts with large surface and moderate curvature. Complex, non-geodesic and concave winding paths can also be fabricated using this technique [7,8]. The tape winding can also be used for making tubular structures such as bicycle frames and satellite launch tubes.

2.3.2   ADVANTAGES

The thermoplastic tape-winding process has the following advantages:

1. The tape-winding process is a clean manufacturing process.
2. The tape-winding process can be used to produce complex structures such as concave surfaces, non-geodesic winding and so on.
3. Thick and large composite structures can be fabricated in a single step.
4. Secondary processing (e.g. oven curing) is not required.

2.3.3   LIMITATIONS

The thermoplastic tape-winding process has the following limitations:

1. The two major requirements of the process are localised heat source and consolidation roller. This renders the process quite complicated.
2. The capital cost of the process is quite high.
3. The raw materials used for the tape-winding process are costly.

2.4   PULTRUSION

The pultrusion is a most cost-effective process for fabricating thermoplastic composite parts. This is a continuous process of fabricating thermoplastic composite products, where the moulding and manufacturing of composites take place in a single step. The dimensional requirements can be achieved by maintaining the quality and integrity of the final product. The reinforced fibres used as raw materials in this process may be in the form of a tape and woven or mat. The basic raw materials used in this process are given in Table 2.3. In pultrusion process, reinforced fibres are drawn from a creel and passed through a thermoplastic resin bath. After impregnating the fibres with resin, it passes through a performing guide system towards a heated die. A pulling system is used to pull the composite through the heated die where the consolidation takes place. The compaction of the composite begins as it passes through the die and reaches the tapered portion of the die. The complete process of thermoplastic pultrusion is shown in Figure 2.2. The speed of the process is regulated by a pulling system that creates a dragging force on the part. The pultruded composite part is cut into the desired length in order to obtain the final part.

The pulling mechanism is an important part of the pultrusion process as it controls the speed of the entire process. The pulling mechanism also ensures the continuous and steady motion of the product at a specified speed. The pulling force that needs to pull the composite through the heating die depends on the resistance forces coming into play during the process. The resistance force is built-up due to the viscosity of the resin, friction inside the cavity, die surface condition, thermal expansion and

TABLE 2.3
Raw Materials Suitable for Processing in Pultrusion Process

Polymers	Polyetheretherketone (PEEK), polyphenylene sulfide (PPS), polypropylene (PP), polyethylenimine (PEI), polyurethane (PU), ISOPLAST, nylon, polyester and so on
Reinforcements	Carbon fibre, glass fibre, aramid fibre, boron and other organic and inorganic fibres and so on

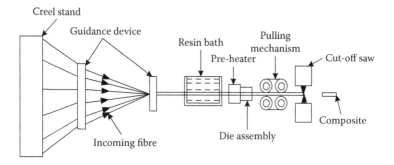

FIGURE 2.2   Schematic of thermoplastic pultrusion process.

complexity of the profile. A higher pulling force is required if the viscosity of the thermoplastic resin is high. Åström [9] reported that the pulling force is directly proportional to the circumference of the composite and slightly less than directly proportional to the pulling speed. Sumerak [10] investigated the influence of different process parameters of thermoplastic pultrusion while fabricating rectangular-shaped glass/polyester composite. The pulling force and the speed recorded during the process were in the range of 1–9.6 kN and 5 mm/s, respectively. The authors have also concluded that the pulling speed highly depends on the matrix formulation, filler content and fibre volume fraction. The other parameters that affect the pulling force and line speed are (1) geometry of the profile, (2) number of parts to be produced, (3) fibre-resin properties and (4) impregnation speed. The thermoplastic pultrusion process can offer an average line speed of 60 m/min. The pulling force highly influences the impregnation of the fibres with resin and the degree of curing. Therefore, pulling force is one of the important factors that decide the quality of the pultruded product. The degree and uniformity of cure can also influence the mechanical properties of the pultruded product [11].

2.4.1 METHODS OF APPLYING HEAT AND PRESSURE

The pressure required for consolidation is supplied by the heated die itself as the pulling system pulls the composite materials through the die. The die is heated to a temperature lower than the melting temperature of the polymer. The heating of die is a challenging task due to the low thermal conductivity of the thermoplastic resin [12]. The temperature requirements for processing of thermoplastic composites are high as compared to the thermoset composites. Therefore, thermoplastic composites are passed through a preheater installed in front of the heated die. The use of preheater also speeds up the process. The common ways to supply the heat to the die are listed in Table 2.4. The selection of heating process, die material and die design are important aspects as they substantially affect the quality of the final pultruded part. The materials generally used in the die are steels as they offer good mechanical properties and suitable for refurbishing. Usually, the surfaces of both upper and lower dies are plated with hard chrome.

TABLE 2.4
Various Sources of Heat Supplied in Thermoplastic Pultrusion

Heat Source	Heating Modes	Remarks
By using platens on both sides of the die	• Electrical resistance • Hot oil • Hot stream	• A common way to heat the die • Good control of the temperature
Radio frequency Microwave heating		Control of heat flux is difficult

2.4.2 APPLICATIONS

The pultrusion process is a simple process that allows fabrication of continuous composites with close dimensional tolerance. The pultruded composites find its various fields of application due to their excellent properties namely higher toughness, recyclability, repairability and high-performance characteristics. The sophistication of the pultrusion method leads to the diversification of size and shape of the profile moulds. In order to meet the complex design of the components, the special *custom-moulded* profiles can be designed. The pultruded composites also provide good structural properties of the product. The various fields of application of pultruded composites are construction, electrical appliances, marine, transportation and sports. The pultruded product can also be used in flooring, fence, bridge, cable tray support member, ladder, transmission pole, tower, wastewater treatment plant, railway, vehicle body panel, hockey stick, ski pole, golf club, arrow, kite and sail mat.

2.4.3 ADVANTAGES

The thermoplastic pultrusion process has more advantage as compared to the thermoset pultrusion process. The various advantages of pultrusion process are listed below:

1. A wide verity of the thermoplastic materials can be used in the pultrusion process.
2. The process is free from styrene emissions.
3. The products fabricated by this process can be easily reformed and repaired.
4. The products produced by this process are light in weight and pose higher strength.
5. The corrosion resistance and electrical insulation properties of the pultruded products are good.
6. The pultruded products are characterised by long-life and low-maintenance requirement.

2.4.4 LIMITATIONS

The thermoplastic pultrusion process has the following limitations:

1. Involvement of high capital cost.
2. Difficult to produce the complex structures.
3. The surface quality of the products is poor.
4. The temperature and pressure required for this process are very high.
5. The raw materials used for thermoplastic pultrusion process are costlier than the thermosetting pultrusion process.

2.5 COMPRESSION MOULDING

The compression moulding is one of the most popular techniques for fabricating thermoplastic composites for commercial purpose. The thermoplastic compression moulding is a high-volume and high-pressure plastic-moulding process. The schematic

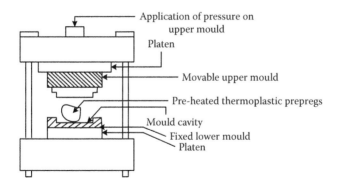

FIGURE 2.3   Schematic of thermoplastic compression moulding.

of the compression moulding is shown in Figure 2.3. The compression-moulding process is suitable for almost all types of reinforcement namely unidirectional fibre, bi-directional or woven fibre mats, randomly oriented fibre mats and short or chopped fibres. The raw materials generally used in compression-moulding process are given in Table 2.5. Natural fibre reinforced with a biodegradable polymer can also be fabricated by this process [13]. Compression moulding is a close moulding process. It has two matched mould halves. The base or lower half of the mould is stationary and the upper half of the mould is movable. The two halves of the mould are closed as the polymer fills the mould cavity. The heat and pressure are applied for a definite period of time as per the requirement of the product to be produced. The melt flows into the mould cavity and takes the shape of the mould. The dimensional accuracy and quality of the end product are dependent on the design of the mould. The curing of the composite takes place either at room temperature or at elevated temperature. After the completion of the solidification process, the mould is opened and the composite product is ejected.

The main controlling parameters of the compression-moulding process are moulding pressure, processing temperature and the duration of application of heat and pressure. Poor interfacial adhesion between the fibre and matrix may occur at low pressure, whereas at high pressure, fibre breakage and expulsion of enough resin may occur. Wakemana et al. [14] studied the hot-flow compression moulding

TABLE 2.5

Raw Materials Suitable for Processing in Compression Moulding

Polymers	Polypropylene (PP), polyethylene (PE), polycarbonate (PC), polyvinyl chloride (PVC), polyetheretherketone (PEEK), acrylonitrile butadiene styrene (ABS), polystyrene (PS), polylactic acid (PLA), polyvinyl alcohol (PVA), nylon, cellulose acetate, soy-based plastic, starch-based polymers and so on
Reinforcements	Glass fibre, carbon fibre, aramid fibre and natural plant fibres (sisal, banana, nettle, hemp, flax etc.)

of glass-mat thermoplastics. It was reported that the minimum pressure required to fill the mould cavity is 50 bars. The moduli of the composites are increased by 3% as the compression pressure is increased from 50 to 200 bars. The time at pressure was found to be a significant parameter affecting the properties of the composites. The modulus was found to be increased by 11% and the void content was reduced by 66% when pressure is applied for 40 s. Giles and Reinhard [15] revealed that compression velocity and the ratio of blank width to the height are the important parameters as compared to the applied pressure. The highest degree of flow was found at a pressure of 137 bars. The temperature and pressure applied during the consolidation process should be carefully chosen. The fibre and matrix properties may change at high temperature, whereas at low temperature, the viscosity of the resin is increased which results in insufficient wetting of fibres with the resin.

2.5.1 METHOD OF APPLYING HEAT AND PRESSURE

The thermoplastic prepregs are heated before transfer to the mould cavity. The heat is applied to the melting temperature of the resin (200°C–230°C) as the prepregs are placed into the mould cavity. Heating can be done either by infrared radiation or by hot air. Curing and solidification of the composite part take place at room temperature or by circulating water at a temperature of 30°C–60°C. The pressure of 70–400 bars is applied by moving the upper half of the mould against the lower half of the mould. Generally, hydraulic mechanism is used to apply the pressure.

2.5.2 APPLICATIONS

This process is applicable for producing various structural components. Bumper beams, recreational vehicle body panels, dashboards, electrical items, knee bolsters, medical equipment, aeroplane parts and other automotive structural components are also manufactured using compression-moulding process. Many consumer items such as office chairs, helmets and so on can also be manufactured by this process.

2.5.3 ADVANTAGES

The compression moulding of the thermoplastic process has the following advantages:

1. Low mould cost.
2. Wastage of the raw material is less.
3. Internal stress and warping are less.
4. The product produced is of high quality and have high-product definition.
5. This process is capable of producing components of varying thicknesses and complexities.
6. The production rate of compression moulding is high.

2.5.4 Limitations

The compression moulding of the thermoplastic process has the following limitations:

1. Higher capital cost.
2. High-fibre volume fraction cannot be achieved.
3. The design of complicated mould with uneven parting lines is difficult.

2.6 INJECTION MOULDING

Injection moulding is a process of manufacturing thermoplastic products in finished form. This is the most suitable and widely used manufacturing process for fabricating fibre-reinforced thermoplastic composites. The schematic of the injection-moulding process is shown in Figure 2.4. This method can be used to fabricate a wide range of dimensions as well as intricately shaped thermoplastic composite products. The raw materials generally used in injection-moulding process are given in Table 2.6. The total time taken by an injection-moulding process is around 20–60 s. The basic steps involved in injection-moulding process are (1) filling, (2) packing-holding and (3) cooling. The thermoplastic polymer is heated in a compression cylinder. Most of the cases the compression cylinder is a reciprocating screw-type barrel that rotates at a speed of 40–80 rpm. Generally, the hydraulic or electric motor is used to drive the reciprocating screw. The melt moves towards the nozzle as the screw starts rotating. The polymer inside the barrel passes through the

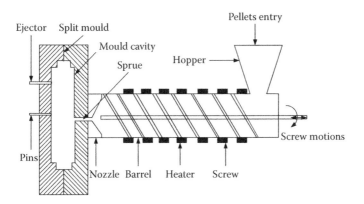

FIGURE 2.4 Schematic of thermoplastic injection moulding.

TABLE 2.6
Raw Materials Suitable for Processing in Injection Moulding

Polymers	Polypropylene (PP), polyethylene terephthalate (PET), nylon, polyester and so on
Reinforcements	Kevlar fibre, glass fibre, carbon fibre and so on

various heating stages. The polymer melts as the temperature reached to its melting temperature. The molten polymer is then forced into the mould cavity at an injection rate of 230 cm³/s and injection pressure of 170 MPa. The mould is designed according to the shape and size of the desired product. The extra material is packed inside the cavity to compensate the shrinkage of the polymer that occurs during cooling. The clamping force of 50–100 MPa is applied to close the mould. During cooling stage, the crystallisation of polymer takes place. The cooling is done below the melting temperature of the polymer (50°C–100°C) by circulating cooled water. Injection-moulded composites take relatively longer time for cooling. Almost 50% of the cycle time of injection moulding is dedicated to cool the mould. The mould is opened at the mould-parting plane and then the final product is ejected using a knockout system. The quality and integrity of the injection-moulded products depend on the back pressure, melting temperature and mould temperature. Many investigations have been carried out to predict the injection pressure and flow-front progression in complex moulds. During filling of the mould cavity, the thermoplastic polymers are subjected to a large amount of shear force, which is proportional to the injection speed. The viscosity of the polymer is increased if the shear rate is in the non-Newtonian region. Hence, the injection speed is set in such a way that the shear rate is in Newtonian region.

The raw material used in injection moulding is mainly in the form of pellets. The fibre-reinforced pellets are formed by passing continuous fibre through a die. The coated fibre is chopped to a length of 10 mm. The length of fibres further decreases as fibre breaks when it passes through the rotating screw and nozzle. The pellets can also be produced by cutting pultruded or extruded composite rod [16–19]. However, the pellets produced from pultruded composite rod may cause damage to the fibres [20–22]. Therefore, short fibre-reinforced thermoplastic composite should be fabricated using direct injection-moulding process instead of using extrusion-injection moulding process. The direct injection-moulding process provides better retention as well as less variation of fibre length [23].

One of the modern technologies of making fibre-reinforced thermoplastic part is water-assisted injection moulding. This technology can produce products with a better surface finish in a shorter cycle time. The shrinkage and warpage of composite are less in this process. The water-assisted injection moulding can be classified into short-shot moulding and full-shot moulding. In short-shot moulding, the mould is partially filled with the melt polymer, whereas in full-shot moulding the mould is fully filled with the melt polymer. Water is injected into the core of the polymer melt as shown in Figure 2.5. The cooling time for water-assisted injection-moulded composites is shorter than the conventional gas-assisted injection-moulded composites [24].

2.6.1 APPLICATIONS

The injection moulding is an ideal moulding method for high-volume production. This method is used to produce various thermoplastic parts that are useful in day-to-day activities such as wire spools, bottle caps, pocket combs, musical instruments,

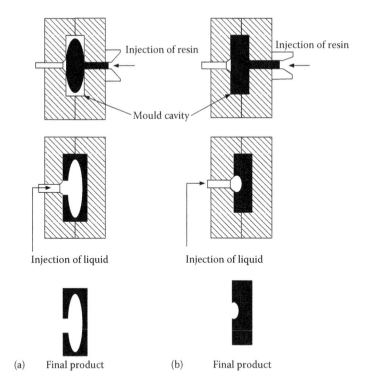

FIGURE 2.5   Water-assisted thermoplastic injection-moulding process: (a) short-shot water-assisted injection-moulding process and (b) full-shot water-assisted injection-moulding process.

chairs, tables, storage containers, mechanical parts (including gears) and so on. The major applications of injection-moulded fibre-reinforced thermoplastic composites are sprockets, computer parts, automotive parts and so on.

2.6.2 ADVANTAGES

The injection-moulding process of the thermoplastic process has many advantages as follows:

1. The complex shapes that are difficult to produce by any other means can be easily manufactured by injection-moulding process.
2. The process is adopted by many industries because of high production rate.
3. The runners, gates and scrap that are used in mould design are recyclable.
4. It can produce a wide range of product with a variety of shapes and size.
5. The products produced by this process are of near-net-shape.

2.6.3 Limitations

The injection-moulding process of the thermoplastic process has the following limitations:

1. High capital investment.
2. This process is not suitable for fabricating low volume parts.
3. This process is not economical for fabricating prototype parts.
4. The quality of the part produced by the injection moulding depends on many factors such as injection pressure, back pressure, melt temperature, mould temperature and shot size. Therefore, the quality of the product cannot be determined immediately.

2.7 AUTOCLAVE PROCESSING

Autoclave processing is a kind of vacuum-bag moulding process as shown in Figure 2.6. Autoclave processing is generally used for manufacturing of various components in the aerospace industry. The process is suitable to fabricate high fibre volume fraction components with any fibre orientation. The raw materials generally used in this process are given in Table 2.7. The thermoplastic prepregs are cut into required dimensions and placed one upon the other as per the required fibre orientation. The stacked thermoplastic prepregs are spot welded together to constrain

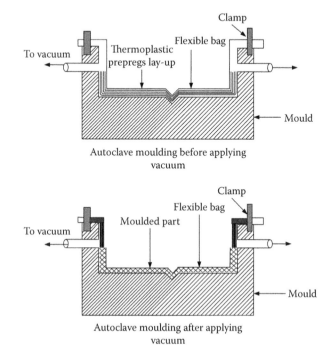

FIGURE 2.6 Schematic of thermoplastic autoclave processing.

TABLE 2.7

Raw Materials Suitable for Processing in Autoclave Moulding

Polymers	Polypropylene (PP), polyethylene terephthalate (PET), polystyrene (PS), acrylonitrile butadiene styrene (ABS), polycarbonate (PC), polyetheretherketone (PEEK), nylon, polyester and so on
Reinforcements	Kevlar fibre, glass fibre, carbon fibre and so on

the relative motion between the layers of prepregs. This is done due to the insufficient tack and drapability characteristics of the thermoplastic prepregs. The entire assembly is then vacuum bagged and placed in an autoclave for consolidation. The vacuum bagging is done in order to remove the entrapped air at the interference of the prepregs and to create an intimate contact between the layers. Bonding between the interfaces of the prepregs begins due to the autohesion process. Due to the internment contact of the adjacent layers of similar thermoplastic prepregs, segments of chain-like molecules diffuse across the interface [25–28]. The extent of molecular diffusion depends on the temperature and duration as given by the Equation 2.1.

$$D_{au} = \chi . \, t_a^{1/4} \tag{2.1}$$

where, t_a is the time elapsed from the start of the autohesion process, and χ is the constant that depends on the temperature (T) given by the Equation 2.2.

$$\chi = \chi_0 \exp\left(-\frac{E}{RT}\right) \tag{2.2}$$

where:

χ_0 is the constant
E is the activation energy
R is the universal gas constant

The mass fraction crystallinity also depends on the value of heat fusion, which is given by the Equation 2.3 [29].

$$\chi_{mc} = \frac{H_M - H_A}{\left(1 - \chi_{mr}\right)H_F^*} \tag{2.3}$$

where:

H_M is the enthalpy of crystal melting
H_A is the enthalpy of additional crystallisation
H_F^* is the theoretical heat of fusion for pure crystalline phase
χ_{mr} is the mass fraction reinforcement

The entire assembly is cooled at a rate of 2°C–10°C/min as the consolidation of the polymer takes place. The crystallisation of polymer takes place during the cooling stage. The vacuum bag is removed and the final product is obtained as the cooling stage is culminated. Some difficulties that arise during autoclave processing are (1) maintaining the accurate fibre orientation, (2) obtaining parts without void and (3) formation of warpage during prepreg lay-up. The processing temperature, pressure and hold-time are the critical process parameters of this process. These parameters should be judicially selected in order to control the dimensional accuracy and temperature fluctuation within the autoclave chamber. Manson and Seferis [29] reported that performance of the autoclave during processing of carbon/PEEK composites was optimum at a temperature range of 375°C–400°C. It was also stated that the consolidation pressure and hold-time have the least effect on the processing of carbon/PEEK composites when the holding pressure and holding time are below 0.2 MPa and 15 min, respectively. If the processing temperature is set without specifying the heating and cooling rates of the processing cycle, the degree of crystallinity may increase to a higher value [29].

2.7.1 METHOD OF APPLYING HEAT AND PRESSURE

The autoclave acts as a pressure vessel. The pressure and temperature required for the process can be maintained by the autoclave itself. The pressure can be applied by two means, that is, by creating a vacuum inside the bag or by applying the external pressure inside the autoclave. The vacuum is created by a vacuum pump, which is attached to the nozzle. The vacuum is maintained at a pressure of 650–750 mmHg. The external pressure of 0.5 MPa is applied externally by injecting pressurised air or nitrogen. When curing is done at high temperature, nitrogen is preferred over the air to avoid burning. The pressurised air and nitrogen also provide the heat required for the process inside the autoclave. The pressure applied by the hot gases results in an increase in the temperature to 390°C. As a result, the viscosity of the resin decreases and easy flow of resin takes place. The cartridge heater can also be placed inside the autoclave to supply heat. Due to the application of heat and pressure, the resin melts and fills the irregular space present at the prepreg interface. The excess resin at the interference of the prepregs layers is also removed due to the application of pressure. The excessive resin may cause poor consolidation. The excess resin may flow normal to the plane of laminate, parallel to the plane of laminate and in both the directions. It was stated that the flow of resin is significantly affected when the fibre volume fraction is more than 65%–70% [30,31].

2.7.2 APPLICATIONS

Autoclave processing is mostly used to produce tough composite parts. Monaghan and Mallon [32] designed and developed a computer-controlled autoclave for sheet forming of thermoplastic composites. Mallon and O'Bradaigh [33] developed a pilot autoclave for diaphragm forming of continuous fibre-reinforced thermoplastic composites. Single-curvature bend over 90° can be successfully formed using the autoclave. The pre-consolidated sheet can be used to manufacture complex composite parts. The use of pre-consolidated sheet decreases the cost of manufacturing, as there is no need of prepreg handling equipment.

2.7.3 ADVANTAGES

The autoclave processing of thermoplastic process has the following advantages:

1. The component with high fibre volume can be fabricated with this process.
2. This process is flexible for producing a component of any fibre orientation.
3. Prototype part can be manufactured by this process.
4. This technology is relatively simple. Tool design is simple for this process.

2.7.4 LIMITATIONS

The autoclave processing of thermoplastic process has the following limitations:

1. Initial capital cost is more.
2. High temperature and pressure are required to process the material.
3. Autoclave processing of thermoplastic composites is difficult in comparison to the autoclave processing of thermosetting composites.

2.8 DIAPHRAGM FORMING

The diaphragm-forming process is not applicable for fabricating thermosetting composites. This process is particularly used to fabricate thermoplastic-based polymer composites. The principle of diaphragm forming is same as that of vacuum forming of thermoplastic sheet. The raw materials generally used to fabricate composites using diaphragm-forming process are given in Table 2.8. A laminate sheet is formed by stacking down the unidirectional prepregs one upon the other and spot welded them together. The prepregs are heated close to the melting temperature of the thermoplastic resin. After heating, the laminate sheet is placed between two thin diaphragms of viscoelastic nature. The diaphragm-forming process is schematically shown in Figure 2.7. The diaphragm materials are usually superplastic aluminium alloys, polyimide (PI) films and UPILEX. The air between the lower diaphragm and female mould is evacuated by creating a vacuum. It is very important to eliminate the air between the diaphragms as the entrapped air results in poor consolidation of the composite part. The composite laminate sheet floats freely between the two diaphragms. The sheet is then clamped around the perimeter. The composite laminates deform plastically inside the mould cavity upon application of either vacuum, pressure, or both. Biaxial tension is maintained during deformation of the laminate. This helps in eliminating the defects such as laminate wrinkling, splitting and thin spots.

TABLE 2.8
Raw Materials Suitable for Processing in Diaphragm Forming

Polymers	Polyphenylene sulfide (PPS), polyetheretherketone (PEEK), polyetherimide (PEI), polyethylene terephthalate (PET), nylon and so on
Reinforcements	Glass fibre, carbon fibre and so on

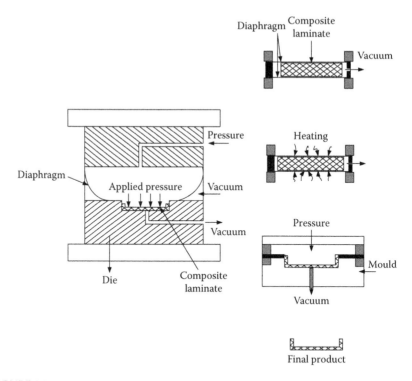

FIGURE 2.7   Schematic of thermoplastic diaphragm forming.

The pressure required for sheet forming of the fibre-reinforced thermoplastic composites is more than 0.4 MPa (60 psi). The part is cooled under application of pressure to cure the composite part and then, the final product is removed from the mould. The effect of forming temperature during diaphragm forming of carbon/PEEK composites has been experimentally investigated [34,35]. The optimum forming temperature range was found to be 360°C–390°C for fabricating single-curvature bend at 90°. Mallon and O'Bradaigh [33] designed and commissioned a polymeric diaphragm-forming process to fabricate continuous fibre-reinforced thermoplastic composite parts. It was found that the polymeric diaphragm forming has the potential to produce high-performance complex parts in short time as compared to the metallic diaphragm. For fabricating high-temperature thermoplastics (PEEK and PEI), the diaphragm is made of PI, whereas for low-temperature thermoplastics (PC, PET and nylon), the diaphragm is made of silicon membranes. The influence of different process parameters such as temperature, pressure and forming rate has been experimentally investigated during diaphragm forming of continuous fibre-reinforced thermoplastic composites. It was concluded that the shear bulking depends on the forming rate and deformation rate of the diaphragm and composites. Delaloye and Niedermeier [36] reported that stacking sequence of prepregs and geometry of the mould has a significant influence on the fibre wrinkling and laminate buckling. The upper diaphragm can slide over the bottom diaphragm even at low forming rate, that is, less than 1 bar/min. The upper diaphragm slides over the bottom diaphragm due to the viscoelastic nature of the diaphragm at a higher forming rate of 10 bar/min. In diaphragm forming, the forming and consolidation take place due to the following: (1) the molten resin slowly passes through the fibre layers, (2) movement of fibres in the transverse or thickness direction due to the applied pressure, (3) intraply sharing and (4) interply slip [37]. The first two mechanisms ensure good consolidation of the part. The interplay sharing ensures the relative movements of the fibres within the ply both in axial and transverse directions. The fourth mechanism helps to slide the plies over each other according to the shape of the die. The relative movement of the adjacent plies is necessary to accommodate the part definition [34].

2.8.1   HEAT AND PRESSURE SUPPLY

The heat is supplied in diaphragm-forming process by two means: (1) oven heating and (2) autoclave heating. In oven heating, the composite laminate along with the diaphragm is placed in an oven. In the oven, the entire assembly is heated, and hence, the temperature of the composite is reached to the melting temperature of the thermoplastic resin. After heating, the assembly is placed inside a female mould. The air between the lower diaphragm and the mould is evacuated by creating a vacuum. The pressure is applied above the upper diaphragm for forming to take place. The solidification phenomenon starts as the composite laminate comes in contact with the female mould. In autoclave heating, the female mould is heated isothermally after placing the composite inside an autoclave. In general, hot gas is used in the autoclave to heat the mould. The hydrostatic pressure is applied above the upper diaphragm. The vacuum is applied in the same way as described earlier. The entire process of forming can be made faster using separate heat sources and pressure vessels. After forming and consolidation are over, the mould is cooled either inside the autoclave or at ambient air. After completing the cooling process, the vacuum between the diaphragms is removed and the cooled part is then taken out.

2.8.2   APPLICATIONS

The diaphragm-forming process has not gained much importance from the commercial point of view. However, it is very useful for making small series of large parts. Therefore, this forming process has found its huge applications in the aerospace industry for making various parts. The complex parts such as helmets, trays and corrugated shapes with excellent surface quality can be fabricated using the diaphragm-forming process [38]. The compliant diaphragms are preferred for making simple shapes, whereas stiffer diaphragms are preferred for making complex shapes.

2.8.3   ADVANTAGES

The diaphragm forming of thermoplastic composites has following advantages:

1. It is a cost-effective process.
2. Diaphragm forming provides the best control over fibre placement among the various sheet-forming processes.

3. The main advantage of diaphragm forming is high-quality consolidation.
4. Better mechanical properties of the final formed product can be achieved.
5. Diaphragm forming offers a better control of the process and allows forming over different lengths of fibres and different fibre–matrix combinations.

2.8.4 LIMITATIONS

The diaphragm forming thermoplastic composites has the following limitations:

1. The fabrication of complex geometries is a difficult task due to the stretching and creeping of the diaphragms.
2. This process has comparatively larger cycle time (usually 30–60 min).
3. The stiffness of the diaphragm creates complexities in obtaining good quality part.

2.9 HOT PRESSING

The hot-press forming technique offers large-scale production of thermoplastic composite components. The entire process of hot-press forming is shown in Figure 2.8. The raw materials used in this process may be in the form of prepreg, which is a mixture of thermoplastic resin and reinforcement. The raw materials generally used to fabricate composites using hot-pressing technique are given in Table 2.9. The prepregs are cut according to the desired dimensions and then stacked together to achieve the desired thickness of the laminate. The prepregs are spot welded around the edges to constrain the relative movement of the prepregs. The laminate is then

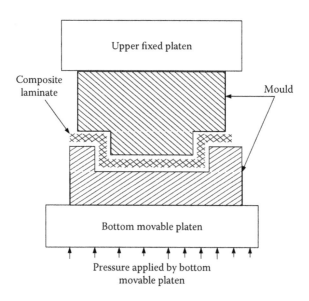

FIGURE 2.8 Schematic of thermoplastic hot pressing.

TABLE 2.9
Raw Materials Suitable for Processing in Hot-Press Forming

Polymers	Polyetheretherketone (PEEK), polypropylene (PP), polyetherimide (PEI), polyethylene terephthalate (PET), polylactic acid (PLA) and so on
Reinforcements	Glass fibre, carbon fibre and natural fibres (jute, flax, hemp etc.)

placed into the mould and the mould is further placed between the heated moulds of compression-moulding equipment. The mould is heated up to a melting temperature of the polymer (380°C–400°C) and then, the pressure is applied. The polymer started to melt and spread throughout the reinforced fibres as the temperature reached the melting temperature of the polymer. A low pressure of 0.3–1 MPa is maintained for 5 min at this equilibrium temperature. The mould is then removed from the compression-moulding equipment and placed in a cold press. The pressure is maintained until the solidification is completely over. If the hot-press equipment has a built-in cooling system, the cooling of the mould takes place in the hot press itself without transferring the mould into the cold press. After solidification is completely over, the mould is opened and the part is removed. The moulds are generally made of stainless steel or aluminium. The moulds are simple and mostly used to fabricate flat laminates. Mould design does not require sprue, runner and gate, as this process is free from injection of raw materials. Releasing agent is applied to the mould surface for easy removal of the composite part. It is important to have intimate contact between adjacent prepreg layers for good consolidation of the final product. The air at the interfaces of the prepregs needs to be removed in order to establish the intimate contact. The autohesion phenomenon takes place at the interfaces due to molecular diffusion. The important parameters that influence the mechanical properties of the composite are temperature, pressure and heating time. Kiran et al. [39] found that the processing temperature of hot-press forming has the most significant effect on evaluating the mechanical properties of jute/polylactic acid (PLA) composites. The tensile and flexural strengths of the composite are found to be significantly affected by the processing pressure and heating time, whereas these parameters are insignificant for impact strength. Placket et al. [40] reported that the tensile properties of the jute/PLA composites are more when it is processed for a lower heating time. The cooling rate, preheating time and recrystallisation-soak time also influence the mechanical properties of the thermoplastic composites [41]. Kumar and Balachandar [42] experimentally investigated the process parameters of hot-press forming while fabricating glass/PP hybrid woven composites. The processing temperature, mould pressure and holding time were found to be the significant parameters affecting the flexural properties of the composite material. The tensile and flexural strengths of the unidirectional kenaf/PLA composites are mostly affected by the processing temperature [43]. Takagi and Asano [44] and Rassmann et al. [45] reported that the tensile strength, flexural strength and modulus are increased with an increase in the moulding pressure. Medina et al. [46] also showed that the mechanical properties of natural fibre-reinforced composites are significantly dependent on the applied pressure.

2.9.1   METHOD OF APPLYING HEAT AND PRESSURE

The heat is applied after the mould is placed between the two heated platens. The heat can be applied electrically. In computerised-controlled hot press, the heating rate, cooling rate and dwell time are controlled by the computer. The top platen is usually fixed and the bottom platen is used to apply the pressure. The amount of pressure applied is measured using a pressure gauge.

2.9.2   APPLICATIONS

The commercial application of hot-press technique is not extensive. The application of hot-press forming is limited to simple components with flat surfaces.

2.9.3   ADVANTAGES

The hot pressing of thermoplastic composites has the following advantages:

1. Hot-press technique allows fabrication of high fibre volume fraction composites.
2. Thermal degradation of natural fibres is less in hot-press process as compared to the injection-moulding process.
3. Unlike injection-moulding process, the fibres in hot-press process are heated only once without any shearing. Hence, the composite part can be achieved with less damaged fibres.
4. Mould design is comparatively easy as the mould is free from injection of raw materials.

2.9.4   LIMITATIONS

The hot pressing of thermoplastic composites has the following limitations:

1. The hot-press technique is not suitable for fabricating complex shapes with thick structure.
2. Only thin and flat plates can be fabricated by this process.
3. Various defects like part distortion, warpage and so on may arise during the fabrication.

2.10   SUMMARY

In this chapter, various process parameters that influence the manufacturing of thermoplastic composite have been adequately discussed. The suitable raw materials for each manufacturing process along with the advantages and limitations of the process have been highlighted. The applications of different manufacturing processes in context of thermoplastic composites have also been presented. The processing parameters of the manufacturing techniques significantly influence the mechanical properties and quality of the end product. Therefore, suitable processing parameters must be carefully selected in order to yield the composite products with the desired quality. The mechanical properties of the final product are significantly influenced by wetting characteristics of fibre, the characteristic of fibre–matrix interface and void content. Therefore, a suitable fabrication process must be judiciously selected for rapid and low-cost production of small- to large-scale parts.

REFERENCES

1. Mazumdar, S.K. 2002. *Composites Manufacturing: Materials, Product, and Process Engineering.* CRC Press LLC, Boca Raton, FL.
2. Mazumdar, S.K. and Hoa, S.V. 1996. Determination of manufacturing conditions for processing PEEK/carbon thermoplastic composites using hot nitrogen gas by tape winding tehnique. *J. Thermoplast. Compos. Mater.* 9(1):35–53.
3. Beyeler, E., Phillips, W., and Guceri, S.I. 1988. Experimental investigation of laser assisted thermoplastic tape consolidation. *J. Thermoplast. Compos. Mater.* 1(1):107–121.
4. Mazumdar, S.K. and Hoa, S.V. 1993. Experimental determination of process parameters for laser assisted processing of PEEK/carbon thermoplastic composites. In *Proceedings of 38th International SAMPE Symposium*, Anaheim, CA, pp. 189–204.
5. Sonmez, F.O. and Hahn, H.T. 1997. Modeling of heat transfer and crystallization in thermoplastic composite tape placement process. *J. Thermoplast. Compos. Mater.* 10(3):198–240.
6. Ghasemi Nejhad, M.N. 1993. Issues related to processability during the manufacture of thermoplastic composites using on-line consolidation techniques. *J. Thermoplast. Compos. Mater.* 6(2):130–146.
7. Coffenberry, B.S., Hauber, D.E., and Cirino, M. 1993. Low cost alternative: in-situ consolidated thermoplastic composite structures. In *Proceedings of 38th International SAMPE Symposium*, Anaheim, CA, pp. 391–405.
8. Wells, G.M. and McAnulty, K.F. 1987. Computer aided filament winding using non-geodesic trajectories. In *Proceedings of 6th International Conference on Composite Mtaerials, ICCM 6 & ECCM*, vol. 1, pp. 161–173.
9. Åström, B.T. and Pipes, R.B. 1993. A modeling approach to thermoplastic pultrusion. I: Formulation of models. *Polym. Compos.* 14(3):173–183.
10. Sumerak, J.E. 1985. Understanding pultrusion process variables for the first time. In *40th Annual Conference*, January 28–February 1, 1985. Atlanta, GA: Reinforced Plastics/Composites Institute, The Society of the Plastics Industry.
11. Jianhua, L., Joshi, S.C., and Lam, Y.C. 2002. Curing optimization for pultruded composite sections. *Compos. Sci. Technol.* 62(3):457–467.
12. Ding, Z., Li, S., and Lee, L.J. 2002. Influence of heat transfer and curing on the quality of pultruded composites II: Modeling and simulation. *Polym. Compos.* 23(5):957–969.
13. Oksman, K., Skrifvars, M., and Selin, J.F. 2003. Natural fibers as reinforcement in polylactic acid (PLA) composites. *Compos. Sci. Technol.* 63(9):1317–1324.
14. Wakemana, M.D., Cain, T.A., Rudd, C.D., Brooks, R., and Long, A.C. 1990. Compression moulding of glass and polypropylene composites for optimised macro- and micro-mechanical properties II. Glass-mat-reinforced thermoplastics. *Compos. Sci. Technol.* 59(5):709–726.
15. Giles, H. and Reinhard, D. 1991. Compression moulding of polypropylene glass composites. In *36th International SAMPE Symposium and Exhibition*, San Diego, CA, pp. 556–570.
16. Baba, B.O. and Ozmen, U. 2015. Preparation and mechanical characterization of chicken feather/PLA composites. *Polym. Compos.* 16:101–113.
17. Bledzki, A.K., Jaszkiewicz, A., and Scherzer, D. 2009. Mechanical properties of PLA composites with man-made cellulose and abaca fibres. *Compos. A* 40(4):404–412.

18. Asaithambi, B., Ganesan, G., and Ananda Kumar, S. 2014. Bio-composites: Development and mechanical characterization of banana/sisal fibre reinforced poly lactic acid (PLA) hybrid composites. *Fibers Polym.* 15(4):847–854.
19. Mechakra, H., Nour, A., Lecheb, S., and Chellil, A. 2015. Mechanical characterizations of composite material with short alfa fibers reinforcement. *Compos. Struct.* 124:152–162.
20. Du, Y., Wu, T., Yan, N., Kortschot, M.T., and Farnood, R. 2014. Fabrication and characterization of fully biodegradable natural fiber-reinforced poly(lactic acid) composites. *Compos. B* 56:717–723.
21. Yao, F., Wu, Q., Lei, Y., Guo, W., and Xu, Y. 2008. Thermal decomposition kinetics of natural fibers: activation energy with dynamic thermogravimetric analysis. *Polym. Degrad. Stab.* 93(1):90–98.
22. Azwa, Z.N., Yousif, B.F., Manalo, A.C., and Karunasena, W. 2013. A review on the degradability of polymeric composites based on natural fibres. *Mater. Des.* 47:424–442.
23. Chaitanya, S. and Singh, I. 2016. Processing of PLA/sisal fiber biocomposites using direct- and extrusion-injection molding. *Mater. Manuf. Process.* 32(6):468–474.
24. Liu, S.J. and Chen, Y.S. 2004. The manufacturing of thermoplastic composite parts by water-assisted injection-molding technology. *Compos. A* 35(2):171–180.
25. Lee, W.I. and Springer, G.S. 1987. A model of the manufacturing process of thermoplastic matrix composites. *J. Compos. Mater.* 21(11):1017–1055.
26. Loos, A.C. and Li, M.C. 1990. Heat transfer analysis of compression molded thermoplastic composites. In *Proceedings of 35th International SAMPE Symposium*, CA, USA: SAMPE, pp. 557–570.
27. Li, M.C. and Loos, A.C. 1990. Autohesion model for thermoplastic composites, center for composite materials and structures. Report No. CCMS-90-03, Virginia Polytechnic Institute and State University, Blacksburg, VA.
28. Dara, P.H. and Loos, A.C. 1985. Thermoplastic matrix composite processing model, center for composite materials and structures. Report No. CCMS-85-10, Virginia Polytechnic Institute and State University, Blacksburg, VA.
29. Manson, J.A.E. and Seferis, J.C. 1989. Autoclave processing of PEEK/carbon fiber composites. *J. Thermoplast. Compos. Mater.* 2(1):34–49.
30. Gutowski, T.G. 1985. A resin flow/fiber deformation model for composites. *SAMPE Q.* 16(4):58–64.
31. Gutowski, T.G., Morigaki, T., and Cai, Z. 1987. The consolidation of laminate composite. *J. Compos. Mater.* 21(2):172–188.
32. Monaghan, M.R. and Mallon, P.J. 1990. Development of a computer controlled autoclave for forming thermoplastic composites. *Compos. Manuf.* 1(1):8–14.
33. Mallon, P.J. and O'Bradaigh, C.M. 1988. Development of a pilot autoclave for polymeric diaphragm forming of continuous fibre-reinforced thermoplastics. *Composites* 19(1):37–47.
34. O'Brátdaigh, C.M. and Mallon, P.J. 1989. Effect of forming temperature on the properties of polymeric diaphragm formed thermoplastic composites. *Compos. Sci. Technol.* 35(3):235–255.
35. O'Bradaigh, C.M., Pipes, R.B., and Mallon, P.J. 1991. Issues in diaphragm forming of continuous fiber reinforced thermoplastic composites. *Polym. Compos.* 12(4):246–256.
36. Delaloye, S. and Niedermeier, M. 1995. Optimization of the diaphragm forming process for continuous fibre-reinforced advanced thermoplastic composites. *Compos. Manuf.* 6(3–4):135–144.
37. Cogswell, F.N. 1987. The processing science of thermoplastic structural composites. *Int. Polym. Proc.* 1(4):157–165.
38. Mallon, P.J., O'Bradaigh, C.M., and Pipes, R.B. 1989. Polymeric diaphragm forming of complex-curvature thermoplastic composite parts. *Composites* 20(1):48–56.

39. Kiran, G.B., Suman, K.N.S., Rao N.M., and Rao, R.U.M. 2011. A study on the influence of hot press forming process parameters on mechanical properties of green composites using Taguchi experimental design. *Int. J. Eng. Sci. Technol.* 3(4):253–263.
40. Placket, D., Anderson, T.L., Pedersen, W.B., and Nielson, L. 2003. Biodegradable composites based on L-Polylactide and jute fibers. *Compos. Sci. Technol.* 63:1287–1296.
41. Rezaei, M., Ebrahimi, N.G., and Shirzad, A. 2006. Study on mechanical properties of UHMWPE/PET composites using robust design. *Iran. Polym. J.* 15(1):3–12.
42. Kumar, B.S. and Balachandar S. 2014. A Study on the Influence of hot press forming process parameters on flexural property of glass/PP based thermoplastic composites using Box-Behnken experimental design. *ISRN Mater. Sci.* 2014:6.
43. Ochi, S. 2008. Mechanical properties of kenaf fibers and kenaf/PLA composites. *Mech. Mater.* 40(4–5):446–452.
44. Takagi, H. and Asano, A. 2008. Effects of processing conditions on flexural properties of cellulose nanofiber reinforced green composites. *Compos. A* 39:685–689.
45. Rassmann, S., Reid, R.G., and Paskaramoorthy, R. 2010. Effects of processing conditions on the mechanical and water absorption properties of resin transfer moulded kenaf fiber reinforced polyester composite laminates. *Compos. A* 41(11):1612–1619.
46. Medina, L., Schledjewski, R., and Schlarb, A.K. 2009. Process related mechanical properties of press molded natural fiber reinforced polymers. *Compos. Sci. Technol.* 69(9):1404–1411.

3 Manufacturing and Characterisation of Shape-Memory Polymers and Composites

Devarshi Kashyap, Charan Mukundan and S. Kanagaraj

CONTENTS

3.1 INTRODUCTION

Shape-memory polymers (SMP) are a class of smart materials that have the ability to be programmed to a temporary shape and return to their original shape on the application of a specific stimulus such as temperature, light, magnetic field, electric field, pH and so on. Though a number of other stimulus methods have been discovered, our primary focus in this chapter is temperature-driven shape-memory effect (SME). The initial shape of the SMP fabricated by suitable manufacturing technique is the permanent/original shape. The secondary/temporary shape is programmed by heating the SMP above its switching temperature (T_{sw}), then deforming it to the required shape and cooling it below T_{sw} as it can be seen in Figure 3.1. The temporary shape is retained by the SMP until it is stored below its T_{sw}. On heating the sample above the T_{sw}, the SMP recovers to its permanent shape. The applications of SMP are in diverse areas such as heat-shrinkable tubes for electronics or films for packaging, implants for minimally invasive surgery, self-deployable sun sails in spacecraft, medical devices or smart fabrics.

SMP has several advantages: (1) easy and flexible programming: the programming for temporary shape can be done through single and multi-step processes; (2) they have a broad range of chemical structures: different approaches for designing

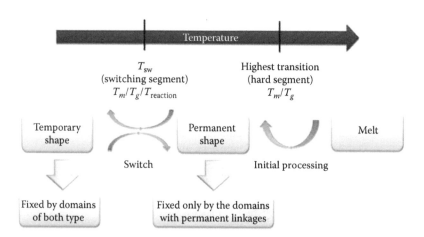

FIGURE 3.1 Schematic representation of the basic principles of the shape-memory effect in polymers. (Reprinted from *Prog. Polym. Sci.*, 49, Hager, M.D. et al., Shape memory polymers: Past, present and future developments, 3–33, Copyright (2015), with permission from Elsevier.)

net-points and switches for various types of SMP; (3) easily tunable properties: the characteristics of SMP can be accurately tuned and engineered very easily using fillers, blending and synthesis methods. The aforementioned advantages and other properties such as biocompatibility, biodegradability, low cost and high recoverable strain make them attractive materials for biomedical applications. Wache et al. (2003) showed the possibility of using SMP as drug-eluting polymeric stents to substitute the currently used metallic stents by improving biological tolerance and reduction of restenosis and thrombosis. The viability of SMP filled with tantalum powder as an embolic agent was investigated by Hampikian et al. (2006). Lendlein and Langer (2002) illustrated the potential of these shape-memory thermoplastics in the form of smart degradable suture for wound closure. Landsman et al. (2017) provided an initial proof of concept for haemostatic sponge of hydrogel-coated SMP foams as a treatment for traumatic haemorrhage. The application of SMP in the medical field is numerous and new applications are being explored. The goal of this chapter is to provide the readers insight into the different manufacturing processes and characterisation techniques for suitable fabrication of SMP composites (SMPC) for biomedical applications.

3.1.1 Mechanism of Shape-Memory Effect in Polymers

The structure of SMP is similar to other types of polymers, which possess complex 3D molecular structures, and they are considered to be the outcome of numerous cross-linking net points and switching segments forming a complex network. The required shape-memory phenomena can be achieved by tailoring two components in SMP structure: firstly, the net points or the hard segments which determine the permanent shape and secondly, the switching segments or soft segments with a suitable T_{sw} to ensure the required shape fixity or temporary shape. The hard segments are formed by chemical cross-linking, crystalline phase, molecular entanglement and interpenetrated network while the switching segments formation can be influenced by crystallisation/melting transition, vitrification/glass transition, liquid crystal anisotropic/isotropic transition, reversible molecule cross-linking and supramolecular association/dissociation.

The SME in the SMP is an entropic phenomenon as discussed by Xie (2011). In its permanent shape, the molecular chains of the SMP are in thermodynamically stable state, that is, highest entropy. When the SMP is heated above the T_{sw}, the chain mobility is increased and the polymer chain conformations are easily changed by applying the deformation load leading to lower entropy state and macroscopic shape change. Subsequently, the temporary shape of the polymer will be fixed by cooling the polymer to a temperature below its switching temperature (T_{sw}), which freezes the molecular chain arrangements, resulting in macroscopic shape fixation. When heat is applied to SMP to reach the temperature above its T_{sw}, the polymer chain mobility is re-activated allowing the SMP to recover its permanent shape.

In this chapter, the different types of manufacturing processes are discussed, which are adopted to process the synthesised polymer into fibres or composites to get desired permanent shapes. Further, we have tried to provide insight into the characterisation techniques in order to have better understanding and control of shape-memory properties.

3.2 MANUFACTURING OF SHAPE-MEMORY FIBRES

SMP can be made into shape-memory fibre (SMF), when it possesses simple or polar side groups, which are orientable, meltable or dissolvable in suitable solvents. The polymers having the required properties are made into fibres by the process called spinning. After synthesis, the SMP in the form of powder or prepregs is dissolved in a solvent or melted to disentangle the polymer chain to make the SMP flowable. The SMP solution is then extruded through a number of holes of desired size and shape in a spinneret and subsequently solidified to get SMF.

The most commonly adopted spinning processes are electrospinning, wet spinning, melt spinning, dry spinning and gel spinning, which are explained below in detail.

3.2.1 ELECTROSPINNING

Electrospinning is one of the widely adopted primary manufacturing processes of SMP due to its low cost, easy and high processing rate. SMP of different diameters can be easily manufactured at a significantly higher rate. In addition, SMP fibres manufactured through electrospinning will possess very fine diameter in the range of 50 nm to 50 μm, high surface area and a very good interconnected network between polymer chains. The electrospun SMP has variety of applications such as textiles, sensors, biomedical engineering and tissue engineering.

A set up of an electrospinning process consists of a syringe connected to a high voltage supply and a collector screen, and it is shown in Figure 3.2. The electrospun SMP is obtained as the result of deformation and squeezing of SMP solution under high voltage and natural drying. In a typical electrospinning technique, SMP solution is taken in a syringe and very high voltage is applied to a syringe tip due to which the hemispherical SMP solution droplets squeezing out of tip undergo severe deformation and transform into conical shape called *Taylor cone*. The SMP droplets will accumulate at the tip and once the critical voltage is reached, which is sufficient enough to overpower the surface tension of the SMP, the elongated fibres are pushed

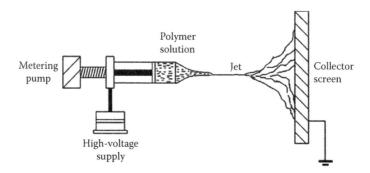

FIGURE 3.2 Electrospinning process. (Reprinted from *Mater. Lett.*, 61(2), Li, X. et al., Preparation of nanosilver particles into sulfonated poly(ether ether ketone) (SPEEK) nanostructures by electrospinning, 421–426, Copyright (2007), with permission from Elsevier.)

towards the grounded collector target. Once the elongated fibres are pushed away from the syringe, the solvent will begin to evaporate, which will be completely dried before reaching the collector. As a result, a continuous fibrous network of SMP is obtained.

In order to obtain SMP fibres with a specific shape, surface morphology, texture and any other properties, a thorough understanding on working parameters of electrospinning is necessary. The parameters such as the voltage applied to the syringe tip, distance of collector from syringe tip, the feeding rate of SMP solution, the solvent used to dissolve the SMP, the viscosity and the conductivity of SMP material are mainly used to tailor the characteristics of the SMP fibres.

Cha et al. (2005) made earlier attempts in electrospinning of shape-memory PU block co-polymers with varying hard segment ratio. Polyurethane-shape memory fibre (PU-SMF) with a range of diameter from 800 to 1300 nm was obtained by tailoring the viscosity of the SMP solution. PU-SMF electrospun at lower viscosity (130–180 cPs) showed a rough surface finish and smaller diameter, whereas a very smooth surface finish and larger diameter fibres were obtained at higher viscosity (530–570 cPs) and the fibres showed a high shape-recovery rate of ~80%. Zhuo et al. (2008) studied the influence of concentration of solvent on the fibre quality during electrospinning process. Different concentrations of N,N-Dimethylformamide (DMF) were used to prepare a Poly(e-caprolactone) diol (PCL)-based SMP solution for electrospinning process. It was observed that the viscosity and the diameter of electrospun fibres increased with the concentration of DMF. For PCL-based PU/DMF solution, 3 and 12 wt.% of DMF in PCL were found to be lower and upper crucial concentration for the electrospinning process, respectively. A thermoresponsive PU-SMP scaffold based on polylactide-co-caprolactone (PDLLA-co-CL) co-polymer was prepared by Tseng et al. (2013) using electrospinning. The SMP was found to have shape-recovery ratio of 94%–96% and a deformation strain-retention capacity of 99%–100%. The ordered fibrous structure obtained by electrospinning enabled the SMF to change shape and internal architecture without any change in the viability of cells present on the scaffold. Another thermoresponsive biomimetic SMP was prepared using electrospinning by Bao et al. (2014) using polylactide-co-trimethylene carbonate (PDLLA-co-TMC). The electospun fibres of PDLLA-co-TMC were obtained in few hundred nanometre with excellent shape-memory properties (>94%) and impressive shape-recovery ratio along with very high-shape fixity ratio (>98%). The very high-surface roughness of PDLLA-co-TMC obtained through electrospinning also improved the osteoblast attachment and bone formation capacities of the SMP.

An electrospun magnetic-active SMP was prepared by Gong et al. (2012) using c-PCL/Fe$_3$O$_4$@CD-M (chemically cross-linked poly [e-caprolactone] reinforced with iron oxide decorated on b-CD grafted MWNT). Excellent alignment of reinforcements was observed along the SMF axis, which resulted in very good heat-inducing SME in the SMF. The Young's modulus and tensile strength of the SMP composite were observed to be 37.6 and 8.5 MPa, respectively. The composite fibres exhibited good shape-memory properties with a shape-recovery ratio of 92% and shape fixity ratio of 83.7%. The activation of SME of the SMP in both hot water and an alternating magnetic field was also demonstrated.

3.2.2 WET-SPINNING SHAPE-MEMORY POLYMER FIBRES

Wet-spinning process produces SMP fibres with good shape-memory characteristics and mechanical properties. Due to low spinning velocity, the output efficiency of wet-spinning process is cumulatively lower than the electrospinning technique. SMP having lower molecular weight processed by wet spinning may result in low breaking elongation and tactility. This type of spinning is applied to polymers, which do not melt and dissolve in non-volatile or thermal unstable solvents.

Wet-spinning technique is similar to weaving process, where many extruded polymeric threads are combined together to form a single, well-aligned SMP fibres. The schematic representation of the wet-spinning process is shown in Figure 3.3. Firstly, the SMP material is made into a solution by dissolving the polymer in a suitably heated solvent. The heated SMP solution is termed as *dope*, which has considerable viscosity to weave out into required fibrous structure. To avoid any voids or irregularities in the extruded SMP fibres, the dope is de-aired to remove any air traps and filtrated to remove impurities before initiating any process. The viscous SMP solution is extruded from the holes of a spinneret into a secondary solution bath called as coagulation bath or spin bath. During this process, the secondary solution diffuses and coagulates into SMP solution thereby precipitating and creating continuous SMP filaments.

The as-spun SMF is moved through godets, where they are washed thoroughly to remove any solvents remaining in the polymer. Subsequently, the fibres are passed through heated bath where the fibres are stretched, so as to orient the polymer chain along the fibre axis. Usually the stretching will range from 20% to 200% based on the desired SMF properties. Finally, the spun SMF is dried and rolled into spools. Heat treatment will also be done to remove any internal stresses stored in SMF during the process of wet spinning.

The spinneret holes are usually 50–100 μm in diameter, which will be decided based on the required quality and texture of the extruded SMP fibres. The properties of SMP can be tailored by modifying the spinning parameters, which facilitates the use of SMP for various applications. The SMP solvent and secondary solutions are different for different polymers and they are chosen based on their chemical reaction kinetics and mechanisms.

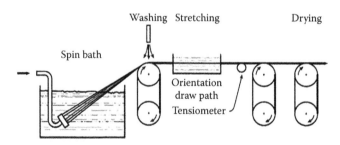

FIGURE 3.3 Wet-spinning process. (Reprinted from *Fibre to Yarn: Filament Yarn Spinning in Textile and Fashion*, Lawrence, C., edited by Sinclair, R., Woodhead Publishing, 213–253, Copyright (2015), with permission from Elsevier.)

Zhu et al. (2006) studied the role of preparation parameters of wet spinning of two different types of SMPU namely PU56-120 and PU66-120 on their SME. The SMPU, which was heated before winding, was having high shape recovery with high recovery stress than the SMPU processed through traditional wet spinning, which was attributed to the well-oriented hard segments obtained by heating before the winding process. Zhu et al. (2007) adopted wet-spinning process to manufacture PU-SMP with different concentrations of hard segments (42%–64%) and observed the effect of post-processing by high-temperature steaming (HTS) on the shape-memory properties of the polymer. The PU-SMF that was post processed by HTS possessed to have the maximum strain, which was increased with decrease in the concentration of hard segments. Also, the HTS processing increased the crystallinity of hard segments from 8.8% to 20.7% in SMF with 64.1% hard segment. The PU-SMF with 42% hard segments and HTS processed possessed to have the maximum strain of 607% compared to 147% observed in unprocessed original polymer. The recovery ratio of HTS processed SMF with 67% hard segments was 90%–95%, whereas the SMF with 55% and 48% of hard segments has almost 100% shape-recovery rate. Ji et al. (2006) studied the effect of concentration of hard segments of wet-spun shape-memory polyurethane (PU) and compared against the SMPU thin films obtained by co-polymerisation technique, where the maximum shape-recovery ratio of SMF was observed to be 92%, which was 7% higher than that of untreated SMPU. In addition, the shape-recovery ratio was observed to be decreased with increase in the concentration of hard segment. The shape fixity ratio of untreated SMPU was observed to be 85%, whereas the same for SMF was found to be 75%.

Cui et al. (2014) fabricated water responsive polyethylene glycol (PEG)-based shape-memory supramolecular hydrogel by wet spinning. The shape-memory properties of SMP such as shape deformation, fixing and recovery were exhibited by improved performance in air and water conditions. The storage modulus of supramolecular hydrogel was observed to be 200 kPa and showed a high stretchability with an elongation at break of 770%. A study on Polyethersulfone (PES)-based SMP by Unger et al. (2005) revealed that PES could be explored as a prospective biomaterial for cell growth. Bouchard et al. (2016) prepared PES/multi-walled carbon nanotubes (MWCNT) shape-memory composite with different concentrations of reinforcement by wet-spinning technique. The mechanical behaviour and electrical percolation of the PES/MWCNT composite were found to be improved after the thermal annealing process. Richardson et al. (2011) studied PU-based shape-memory composites with MWCNT and cellulose nanofibres and reported that the tensile modulus, strength and elongation at break of the films were increased with cellulose fibres content.

3.2.3   MELT-SPINNING SHAPE-MEMORY POLYMER FIBRES

Melt spinning can be employed to the SMP, which can melt and do not degrade during heating. Figure 3.4 shows the schematic diagram of melt-spinning process. The dried SMP is passed into a hopper, which consists of series of heating zones where the polymer is melted above its melting point (T_m). The melted solution is subsequently pushed through spinning line using a pump to control the feed rate of molten SMP. The spinning line consists of a stack of metal disks through which the

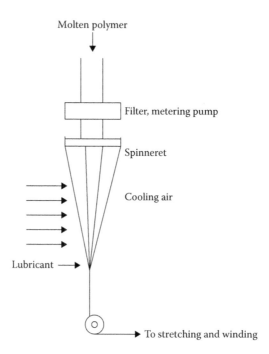

Molten polymer

Filter, metering pump

Spinneret

Cooling air

Lubricant

To stretching and winding

FIGURE 3.4 Melt-spinning process. (Reprinted from *Biotextiles as Medical Implants*, Gupta, B.S., Manufacture, types and properties of biotextiles for medical applications, in edited by King, M.W., Gupta, B.S., and Guidoin, R., Woodhead Publishing, Copyright (2013), with permission from Elsevier.)

melted SMP solution is passed through until it reached the final plate called spinneret. A typical spinneret consists of 100–5000 holes per disk. The size and number of holes on spinneret are decided based on the required properties of extruded SMF. The extruded polymers are cooled, stretched and collected in the form of continuous spools. Stretching is the result of pulling semi-molten filaments coming out of spinneret holes at a higher rate than the flow rate of melted SMP through the spinneret holes. The ratio of the former to the latter speed is called as draw-down ratio.

Meng et al. (2007a) fabricated shape-memory PU fibres by melt and wet-spinning technique and studied the effect of spinning technique on the shape-memory properties of PU fibres. PU-SMF fabricated by melt spinning was found to have improved SME, higher tenacity and breaking strain in comparison to that of SMF fabricated by wet spinning. The improved properties in melt spun PU-SMF were the effect of better soft-segment and hard-segment crystallisation due to higher phase separation. Meng et al. (2007b) prepared PU-MWCNT composites through melt spinning with different concentrations of reinforcement, which played an important role in spinnability and soft-segment crystallisation. At lower concentrations of MWCNT (till 5 wt.%), the spinnablity was improved significantly and it became stagnant and then deteriorated beyond 7 wt.% concentration of MWCNT. An addition of 1 wt.% of MWCNT to PU, the fibre tenacity was improved by 30% and the shape-recovery ratio was increased from 83% to 91%. The reinforcement acted as a nucleating agent till 1 wt.%

and promoted crystallinity of the composites, which was gradually reduced to 8.79% from 16.74% at 7 wt.%. The presence of MWCNT and its interaction with SMP in the composites helped to store the internal elastic energy during loading and shape fixing, which improved the recovery force. Meng and Hu (2008) manufactured PU/MWCNT SMP composite using melt spinning and obtained perfectly aligned MWCNT along the fibre axis of PU fibres. The maximum shape-memory fixity and shape-recovery ratio of melt spun composite were observed to be 86.3% and 98.6%, respectively, whereas the corresponding values for wet-spun composite were observed to be 81.5% and 94.22%, which is due to well-aligned MWCNT in melt-spun composites in comparison to that of wet-spun composites. MWCNT reinforcement helped in storing and releasing the elastic energy during stretching and shape-memory phenomenon.

3.2.4 DRY SPINNING

Dry-spinning technique, as shown in Figure 3.5, is preferred for non-meltable SMP, which cannot be processed through any of the previously discussed methods. The SMP manufactured through dry-spinning process is dissolved in a suitable volatile solvent. Solvents with low boiling point, high volatility and evaporation rate are preferred to achieve high manufacturing rate. The viscous SMP solution is extruded through spinneret using a controlled pump at required flow rate. Prior to extrusion, the solution is filtered to remove any undissolved polymer particles to avoid blockage in the spinneret holes. The extruded viscous polymeric filaments are passed through a drying chamber where hot air is passed to evaporate solvents and purged out continuously. As a result, the filament fuse together to form a single thick SMF, which is stretched by drawing followed by drying. The removal of any solvents present after drying is ensured by subsequent washing. The SMF is then dried and packed. Though the dry-spinning process is faster than wet spinning, the total length of SMP spool that can be manufactured is comparatively lower than the former method,

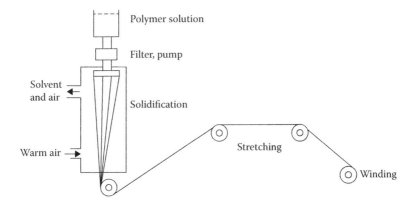

FIGURE 3.5 Dry-spinning process. (Reprinted from *Biotextiles as Medical Implants*, Gupta, B.S., Manufacture, types and properties of biotextiles for medical applications, in edited by King, M.W., Gupta, B.S., and Guidoin, R., Woodhead Publishing, Copyright (2013), with permission from Elsevier.)

which is the result of chemical steps and solvent evaporation that are involved in the process. Thus, the process is expensive and it is desirable, when texture and quality are very crucial parameters for the specific applications. Hu et al. (2009) fabricated SMF by dry-spinning process and it was concluded that formation of urea–urethane groups resulted in high mechanical properties fibres with good heat stability.

3.2.5 GEL SPINNING

Gel-spinning technique is adopted to manufacture SMF from high molecular weight (M_w) SMP. Due to the high M_w, the polymers need very high mixing rate to completely dissolve in a solvent. The SMP is passed through two mixing chambers to ensure complete dissolution and low viscosity suitable for extrusion and the schematic diagram of the process is shown in Figure 3.6. In the first mixing chamber, the SMP is mixed with high temperature, non-volatile solvent and stirred for uniform distribution. After adequate stirring, the solution is transferred to the intense mixing vessel where the SMP solution is vigorously agitated to ensure the complete dissolution of the polymer. The gel-like solution is then passed into a barrel, which is connected to a motor pump, which maintains a uniform flow of the SMP solution horizontally from vessel outlet to the barrel outlet. Another motor attached at the barrel outlet will impart essential force to extrude the SMP solution through the spinneret at a required flow rate. The SMP gel is extruded into the quenching bath

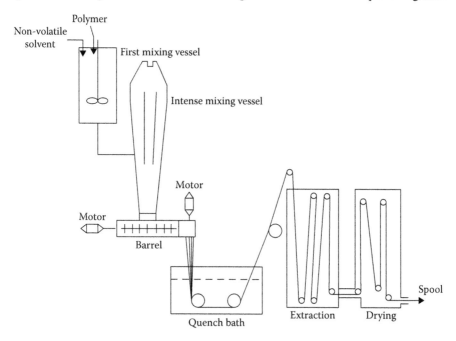

FIGURE 3.6 Gel-spinning process. (Reprinted from *Biotextiles as Medical Implants*, Gupta, B.S., Manufacture, types and properties of biotextiles for medical applications, in edited by King, M.W., Gupta, B.S., and Guidoin, R., Woodhead Publishing, Copyright (2013), with permission from Elsevier.)

that forms SMF with few entanglements and disorientation. The SMF is oriented to its axis by stretching and drawing in an extraction chamber at a very high rate (~30x) than the extrusion rate followed by drying and spooling.

Maksimkin et al. (2014) manufactured SMP with very high M_w such as ultrahigh molecular weight polyethylene (UHMWPE) by gel spinning. Litvinov et al. (2011) studied the changes in morphology of UHMWPE and its molecular-scale phenomena during the drawing of fibres by gel-spinning technique. It was found that the fibres are crystalline and contain a very small amount (~1%) of amorphous phase, which were due to the nanovoids produced during the drawing process. However, it was realised that gel-spinning technique produced near perfect single crystal structure fibre morphology and fibres highly oriented along the axis. Xu et al. (2016) fabricated UHMWPE fibres by gel spinning expected some shish-kebab structure in their spun fibres, which were due to lower drawing speed. By increasing the drawing ratio and time, the crystallinity of the fibres was significantly improved. Due to the high crystallinity and very good orientation of fibres, the gel-spun SMF has very high strength than that of the fibres produced by conventional melt and wet spinning. These results suggest the viability of the use of highly oriented, crystalline UHMWPE SMF as a potential SMP. Gel-spinning method is also used to manufacture shape-memory hydrogel, which has potential biomaterial applications. Lu et al. (2014c) prepared hydrogel with covalently cross-linked polyacrylamide (PAAm) and ionically cross-linked carrageenan by gel-spinning technique. The hydrogel SMP showed the fracture energy and strechability of 9500 J m^{-2} and 1710%, respectively.

3.3 MANUFACTURING OF SHAPE-MEMORY POLYMER AND COMPOSITES

Shape-memory polymer composites (SMPC) consist of a SMP reinforced with particles and/or fibres to tailor its properties. The reinforcements having good magnetic, thermal or electrical properties are used, which will help to apply stimuli to the SMP. In addition, the reinforcement may also need to have better thermal stability and toughness to improve the SMPC properties. The SMP in pellet or powder form is mixed with reinforcement and is manufactured into permanent shape using moulding, extrusion and hand lay-up process.

3.3.1 INJECTION MOULDING

Injection moulding is one of the widely adopted secondary manufacturing processes for the mass production of SMP objects. This process is excellently suitable to fabricate complex geometries or shapes in shorter time and with low rejection rate. Despite its significant principal cost, the durability of the process is high and the cost per object at a higher rate is very low.

In injection moulding, the pelletised SMP is fed into *feed hopper* that sends the SMP to the feed zone. It pushes the SMP to the compression zone at the required feed rate, where the SMP is heated to a semi-solid state and compressed by the extrusion screw such that no voids or air gap present among SMP bulk. The compressed SMP is then pushed into *metering zone*, which is surrounded by heating elements that gradually heats the semi-solid SMP to little higher temperature than its melting temperature (T_m). The melted polymer solution is pushed at high pressure into the die blocks from the barrel, which has a nozzle construction. Due to the high pressure at the exit of the barrel, the molten polymer will flow continuously with no voids even at high temperature. The molten polymer filled in the cavity of die blocks is heated for required time and allowed to cool to room temperature using the external cooling system. Finally, injection-moulded SMP object is removed from the die blocks.

Ni et al. (2007) fabricated carbon nanotube (CNT)-reinforced SMPU by injection moulding with a homogenous dispersion of CNT in the polymer. The uniform dispersion resulted in high yield strength and Young's modulus. The composite also exhibited high shape recovery of nearly 90% even after several cycles. A small addition of CNT of about 3.3 wt.% increased the recovery stress in SMPU by two folds than the pure SMPU. Oyama [2009] manufactured a polylactic acid (PLA) co-polymer reinforced with poly(ethyleneglycidyl methacrylate) (EGMA) by extrusion and injection-moulding process. She reported on the effect of low and high molecular weight PLA namely Lacea H-100 and Lacea H-400, respectively and the annealing process on the properties of the SMP. The addition of EGMA improved the crystallinity and increased toughening and impact strength without affecting the heat resistance of the polymer. The SMP composites reinforced with H-100 found to have high flexural modulus than the polymer reinforced with H-400. The annealing process improved the impact strength of SMP significantly irrespective of reinforcement concentrations.

3.3.2 COMPRESSION MOULDING

The die used in compression moulding is a two-part system that comprises upper and lower dies, which are held together into which the polymer will be melted and compressed. The dies are attached with heating platens, which supply the necessary heat for the process. The polymer is placed in the die, which is heated little above the melting point of the polymer and required pressure is applied. Once the reaction is over, the die is cooled and the object is ejected. In order to provide a smooth removal process of moulded objects, an ejection system consists of ejector actuator, ejector plate and the pins attached to the lower die.

Kunzelman et al. (2008) prepared a photomechanical SMP using poly(cyclooctene) (PCO) cross-linked with oligo (p-phenylene vinylene) using compression-moulding process. The SMP exhibited reversible temperature-sensing capabilities for temperature above T_m. Under UV illumination, with change in temperature, the appearance of the SMP composite changed from green (524 nm) to red (620 nm) wavelength range. Also, the SMP composite exhibited an excellent one-way shape-memory behaviour.

Mendez et al. (2011) prepared water-activated shape-memory composites using rubbery PU matrix with various concentrations of rigid cotton cellulose nanowhiskers (CNW) by compression-moulding process. The presence of hydrophilic cellulose led to water uptake and aqueous swelling behaviour, which was increased linearly with the concentration of CNW. The shape fixity of PU polymer was increased from 13.3% to 60.7% and 74.3% with the addition of 10% and 20% of CNW, respectively. Correspondingly, the shape recovery of PU polymer was increased from 1.6% to 44% and 55.2% for 10% and 20% of CNW-PU composites. For the SMP, which exhibit very high melt viscosity at high temperature such as sulfonated EPDM, Makowski and Lundburg (1990) used stearic acid derivatives as a plasticiser to make them suitable for compression moulding. Weiss et al. (2008) added zinc stearate (ZnSt) as a plasticiser to UHMWPE. The presence of ZnSr reduced the melt viscosity of UHMWPE and made the polymer composite suitable for compression moulding. With the addition of 33.3 wt.% ZnSt, the composite was stable and clear and the shape-recovery ratio was found to be more than 90% for many trials. The addition of plasticiser significantly influenced the critical temperatures (T_c) of the SMP composites.

3.3.3 Extrusion

Extrusion is a continuous process where thermoplastic polymers are melted and re-shaped with the help of an end die. The extrusion machine consists of a screw and a number of heating zones. The helix angle and pitch of the screw are the key factors that determine the output of the polymer at a constant speed. The polymer is fed into the extruder from the hopper and it reaches the metering zone after melting, which is surrounded with heaters to produce homogenous polymer melt with suitable flow properties. The metering zone streamlines the flow of polymer from pulsating to a uniform flow to the mandrel. The heat produced in the metering zone is the combined effect of friction due to shearing of polymer by screw and the external heating of zone by electric source. The molten SMP melt is finally extruded by mandrel through an end plate die that decides the geometry of the extruded polymer. The extruded polymer is cooled to room temperature and solidified with the help of a cooling chamber.

Defize et al. (2011) prepared a poly (e-caprolactones)-based thermoreversible SMP composite using a co-extruder. The SMP composite with Furan-PCL and Maleimide-PCL exhibited excellent shape-memory properties, which was attributed due to homogeneous blending of two functional polymers during the extrusion process. The proper blending of two functional polymers resulted in near 100% fixity ratio and 99% of recovery ratio. To study the effect of moisture content on the shape-memory properties of PU, Yang et al. (2006) processed the SMP by extrusion process and tested. It was noted that bound water absorbed by polymer significantly reduced the glass-transition temperature and tensile behaviour of SMP. Ahir et al. (2006) demonstrated the effects of extrusion and drawing alignment on the SME of ether-based PU. The photo-stimulated SMP processed by extrusion exhibited a very good nematic alignment, which increased the tensile strength and reversible SME significantly. Hampikian et al. (2006) fabricated a SMP-based aneurysm coil with 3 vol.% of tantalum radiopaque material using a customised extrusion process. The SMP deployed in an aneurysm model, which was programmed into a helical configuration, showed a satisfactory shape-memory properties and was unaffected by the hydrodynamic forces exerted by the hot fluid. The SMP processed by extrusion process can be used in a range of application from temperature sensing to biomedical applications.

3.3.4 Resin Transfer Moulding

Resin transfer moulding (RTM) is a vacuum-assisted technique and is widely adopted technique to manufacture thermoset SMP. RTM is a four-stage process as illustrated

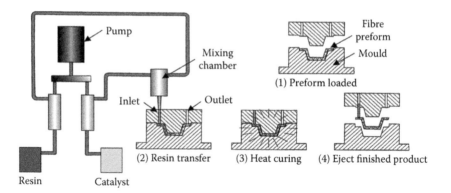

FIGURE 3.7 Step-by-step process of resin transfer moulding.

in Figure 3.7 where the SMP polymer and its hardner/catalyst is compressed into a preform and heated to get the required objects.

Firstly, a preform (fibre or glass) is kept in the mould and clamped tightly. The top mould containing inlet and outlet is attached to the resin mixing chamber. The resin and catalyst are mixed vigorously in the mixing chamber to initiate the chemical reaction between them. The mixed solution is pushed into the tightly held mould through inlet valve and is filled through the cavity in the preform. With the applied pressure during mould filling, the air entrapped in resin-catalyst mixture is removed through outlet valve. The mould is heated to accelerate the curing reaction until complete curing is ensured. Finally, the finished product is cooled and ejected from the moulds. The product obtained from RTM possesses high strength-to-weight ratio and also cheaper than compression-moulding process.

The reinforcements in SMP are dispersed using high-energy sources such as ultrasonic waves from various sources. Solvents such as acetone and ethanol are used to reduce the viscosity of high viscous resin so as to enable homogenous dispersion of reinforcement in the SMP resin. Once dispersion of reinforcement is achieved, subsequently the solvent is removed from resin through evaporation. Sonication bath and tip sonicator are commonly used ultrasonic source for dispersion.

Lu et al. (2010) prepared an epoxy/CNF shape-memory composite with the incorporation of a carbon nanopaper. The introduction of lower concentration of carbon reinforcements reduced the resistivity of SMP composite by 10^{16} times than the pure SMP resin. The reduction in resistivity improved the electrical resistive heating characteristics of the composite. As a result, the shape-recovery time was significantly improved. The SMP composite showed nearly one-fold increase in shape-recovery time from 361 s to 186 s but the shape-recovery ratio was reduced from 98% to 83% with a gradual increase in concentration of CNF in the composite. Lu et al. (2011a) manufactured an electroactive styrene-based SMP with sub-micron nickel strands and self-assembled MWCNT reinforcements, which exhibited excellent electrical and thermal properties. The RTM process helped in achieving homogeneous dispersion of reinforcement and perfect compaction of composite, which resulted in outstanding and optimised SME in the presence of electric signals. Lu et al. (2011b) prepared an electrically responsive epoxy/CNT nanopaper using styrene-based resin and MWCNT nanopaper. The superior bonding between nanopaper and resin improved the shape-memory characteristics of SMP. Lu et al. (2014a) fabricated an epoxy-based SMP composite reinforced with high conductive boron nitride and CNT. The shape-memory characteristics of SMP composite under IR-induced heating showed an evenly distributed CNT and boron nitride and the response of SMP composite to the heating was proportional to its concentration in the composite. Also the dual reinforcement improved mechanical properties and glass-transition temperature of the composites with increase in concentration. The combined effect of reinforcement in transforming IR light into heat and induce SME resulted in excellent IR-responsive SMP composite. Lu et al. (2014b) prepared a Joule heating triggered PU/MWCNT shape-memory composite with multi-layer of carbon nanopaper. The composite prepared by RTM process possessed to have a good interfacial bond between nanopaper and the polymer interface. The self-assembled and multi-layered nanopaper created a continuous conductive path for electrical and thermal stimulus throughout the composite. The increase in nanopaper layers from one to four reduced the resistivity of the composite from 6.41 to 1.67 Ωcm, respectively.

3.3.5 HAND LAY-UP PROCESS

Hand lay-up is considered to be the most economical, easier and simple manufacturing process. This process is mostly adopted to manufacture thermoset polymers with simple shape and less intricate details. Hand lay-up is a time-consuming, slow process as no external pressure is applied that can speed up the reaction rate.

One part of polymer and reinforcement mixture and another part of hardener/catalyst are mixed mechanically by stirring to initiate the chemical reaction between resin and the hardener. Once the reaction is initiated, the solution is poured into the mould and rolled as illustrated in Figure 3.8. During rolling, the air bubbles entrapped in the composite are removed. Followed by rolling, the surface of composite is smoothed using a brush and allowed to cure. After the curing process, sample is ejected from the mould and processed. The rate of curing of polymers will change with temperature that will also affect the mechanical and shape-memory properties of the polymer.

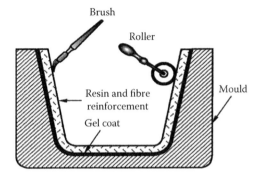

FIGURE 3.8 Hand lay-up process.

Nji and Li (2010a) prepared a shape-memory composite with self-healing mechanism using co-polyester-reinforced polystyrene (CP-PSMP) by hand lay-up process. The addition of 6 vol.% of CP to PSMP helped in achieving a novel biomimic two-step close-then-heal (CTH) scheme with a healing efficiency of 65%. This CTH polymer can be used in healing of structural-length scale damage. Nji and Li (2010b) fabricated a 3D woven shape-memory composite fabric using polystyrene (PS) reinforced with glass micro-balloons and E-glass fabric by gravity-assisted method. The composite SMP was then poured into a mould and cured. The healing response of the SMP composite to various impact strength was analysed. The continuous impact energy test under different levels of impact energy revealed that the impact energy has significant effect on the healing efficiency of the 3D woven fabric. When a level I (32J) of impact energy was applied, the 3D fabric perforated at 9th impact, whereas the same polymer perforated at 5th impact itself under level II (45J) impact energy. The change in perforation limit was due to the applied impact energy, which affected the healing efficiency of the SMP composites. Similarly, 3D fabric that was allowed to heal after each impact showed an increased limit to perforation. Under level I and level II of impact energy, the fabric perforated at 15th and 7th impact, respectively.

3.4 CHARACTERISATION OF SHAPE-MEMORY POLYMERS

The synthesised SMP and their composites are tested and characterised for better understanding of their nature and suitability for end applications. This section discusses about the methodology followed to characterise the shape-memory behaviour of the test materials. Some of the common techniques for traditional polymer characterisation are useful tools for studying SMP. Techniques such as Fourier transform infrared spectroscopy (FTIR), nuclear magnetic resonance (NMR) spectroscopy, wide-angle X-ray diffraction (WAXD) and so on are invaluable tools to study the structural and chemical characteristics of a polymer. Thermal techniques such as differential scanning calorimetry (DSC), thermo gravimetric analysis (TGA) and dynamic mechanical analysis (DMA) provide valuable insight into the thermal properties of a polymer. The morphology of the polymer can be studied by scanning electron microscope (SEM), transmission electron microscopy (TEM) and so on. SME is characterised with the help of thermomechanical cyclic test or bent test using a universal testing machine (UTM) or DMA. The above-mentioned techniques are discussed below to familiarise the reader with the techniques and help them to choose the correct tools for their characterisation.

3.4.1 Mechanical Testing

The mechanical properties of the polymers are tested by simple tensile or compression testing in a UTM, where the sample is subjected to a constant strain rate along a single axis until failure at constant temperature. The parameters obtained from the study are ultimate tensile strength, Young's modulus, yield strength, modulus of resilience, toughness, elongation at yield and elongation at failure. Compression test is done to characterise the materials that cannot withstand tensile loading such as gels and extremely hydrophilic polymers. In addition, uniaxial tensile tests are recommended for isotropic materials, whereas anisotropic samples are required to be tested using biaxial tensile systems. Thermal-transition temperature of the SMP should be taken into consideration and the polymer should be tested at a suitable temperature in order to ensure its usage for the specific application.

SMP is generally weak for load carrying applications, and thus different attempts were made to increase its strength by using suitable reinforcements. Ohki et al. (2004) carried out cyclic mechanical test on the SMP MM4510 (Diaplex Co. Ltd.) reinforced with glass fibres at 25°C to study the effect of cyclic stresses on the composite. Further, static tensile test was performed at glass-transition temperature (T_g) of the SMP, that is 45°C and at $T_g \pm 20$°C. It was reported that the higher filler content increased the tensile strength and Young's modulus of the composite and the elongation at failure was decreased. The maximum tensile strength of 10, 20 and 30 wt.% of composite specimen below T_g was observed to be 9%, 23% and 44% higher than that of pure specimen, respectively. An increment of 8%–15% in comparison with bulk specimen at the higher temperature of $T = 45$°C and 65°C (T_g and $T_g + 20$ K) was reported. The elastic modulus of the SMPC at temperature below T_g was observed to be 50 times greater than above T_g. It was concluded that the potential elasticity in crystalline region and glassy state in amorphous region increased the modulus below T_g, whereas the micro-Brownian movement of amorphous phase above T_g decreased the modulus. The composites showed improved resistance to cyclic loading and crack propagation increase in filler content.

DMA was used by Bao et al. (2016) for studying the tensile properties of electrospun HAp/PLMC composites at 37°C. The stress–strain graph was obtained by increasing the stress from 0 at a stress rate of 0.5 MPa s^{-1} and recording the corresponding strain. A 300% and 500% increase in tensile strength of the 1 and 2 vol.% Hap/PLMC was reported. The reinforcing effect of HAp nanoparticles and the interaction between the functional groups of HAp and PLMC were attributed for the marked increase in mechanical strength. However, the strength was decreased due to agglomeration of HAp at higher concentrations, which disrupted the fibre chains that lead to accelerated breakage of nanofibres during the tensile test.

3.4.2 Thermal Characterisation

3.4.2.1 Dynamic Mechanical Analysis

The DMA is a powerful analysis tool to study the viscoelastic properties of polymers. DMA is widely used to characterise the properties of a material as a function of frequency, temperature, time, strain, stress, atmosphere or a combination of these parameters. Unlike traditional tensile testing, in a DMA test, an oscillating force is applied across a range of temperature and frequency and the resulting response is recorded and reported as stiffness and damping. The sample can be tested in tension, compression and shear or bending depending on the requirement of the end applications. Frequency and temperature sweeps are the most common modes for DMA. The phase lag between the applied stress and measured strain reflects the material's tendency to flow (viscosity), whereas the sample recovery reflects the material's stiffness (modulus), which can be obtained in the form of storage modulus E′,

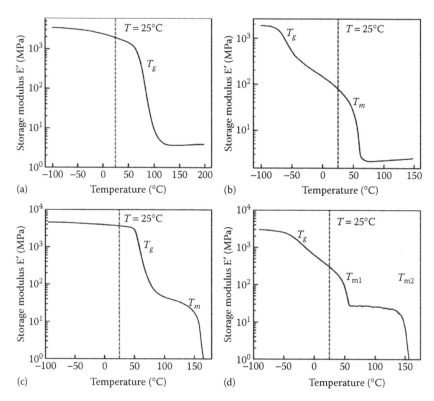

FIGURE 3.9   Tensile storage modulus versus temperature curves: (a) chemically cross-linked glassy thermosets, (b) chemically cross-linked semi-crystalline rubbers, (c) physically cross-linked thermoplastics, and (d) physically cross-linked block co-polymers. (Liu, C. et al., *J. Mater. Chem.*, 17, 1543–1558, 2007. Reproduced by permission of The Royal Society of Chemistry.)

loss modulus, E″ or tan delta curves. The storage modulus reveals the material's ability to store energy, whereas the loss modulus reveals viscous energy loss. DMA also measures the glass-transition temperature of polymers, which is critical in the characterisation of semi-crystalline SMP.

DMA not only provides the glass transition and the melt temperature, but also other transitions that occur in the glassy or rubbery plateau. The storage modulus versus temperature curve for different SMP is shown in Figure 3.9, where the influence of thermosets, thermoplastics, semi-crystalline rubber and co-polymer can be seen clearly.

The primary shape of covalently cross-linked glassy thermoset networks of SMP is covalently fixed, hence once processed, they cannot be re-shaped. This type of SMP has a sharp T_g and it is the switching temperature of the SMP. Amorphous co-polyester–urethane networks (Alteheld et al. 2005) and cross-linked ester-type chemical cross-linking, chemically cross-linked polycyclooctene (PCO) (Liu et al. 2002) and poly(ε-caprolactone) cross-linked by γ-radiation (Zhu et al. 2003) are the few examples of the category. Physically cross-linked glassy co-polymers can be easily processed with conventional thermoplastic fabrication techniques. The material flows easily when the processing temperature surpasses the T_m or T_g of the hard segment. The T_m or T_g of the soft segments acts as the switching segments. Multiblock PU with PEO as soft segments synthesised by Korley et al. (2006) is an example of physically cross-linked semi-crystalline block co-polymers.

DMA can be further used for SME characterisation, which has been discussed later in the chapter.

3.4.2.2 Differential Scanning Calorimetry

The calorimeter detects the change in heat flow in a material with reference to a known sample while maintaining the same temperature. DSC is a useful technique for studying glass-transition temperature, melting point, crystallisation temperature, reaction kinetics, oxidative/thermal stability, specific heat, curing time, crystallisation time, rate and degree of polymerisation of polymers.

DSC is primarily used to study the switching temperature of the shape-memory polymers. The glass-transition temperature and the melting temperature of the shape-memory polymers are observed as step transition (second-order transition) and a peak (first-order transition), respectively, and it is shown in Figure 3.10. The mid-point of the onset and end temperature of the step transition is usually reported as the glass-transition temperature and the melting peak is represented as the melting temperature. The crystallisation peak is observed above T_g, the highly mobile polymer chains gain enough energy to form ordered arrangements and undergo crystallisation.

Though DSC is a very resourceful technique, it is not ideally suitable to investigate all the mentioned properties. DSC can reveal polymer degradation if melting temperature is lower than expected; however, TGA is a better technique to study degradation. Likewise, DMA is a better tool to detect sub-T_g thermal transitions. The decrease

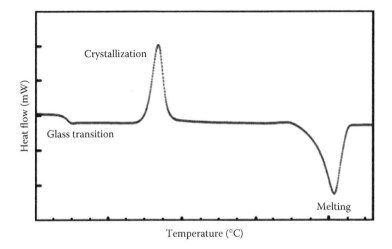

FIGURE 3.10 A typical DSC curve for a polymer that undergoes glass transition, crystallisation and melting.

in T_g of the PU SMP due to moisture was reported by Yang et al. (2005). They observed that the T_g was decreased remarkably with immersion time. They concluded that the moisture acted as a plasticiser, improving the polymer network mobility that decreased the T_g. Hasan et al. (2016a) reported the dry T_g and wet T_g of the nanotungsten-loaded aliphatic PU/urea SMP foams developed for embolic occlusion using the DSC. They found that the T_g was increased by 10°C for the dry foams and by 7°C for the wet foams. The nanoparticles restricted the polymer mobility at the molecular level and increased the physical cross-link within the SMP leading to increased T_g. Nji and Li (2010a) used DSC to confirm the compatibility between polymer–polymer interfaces of polymer composite. They predicted good compatibility between veriflex and co-polymer co-polyster by the single glass-transition criterion. The glass-transition temperature of the composite ($T_g = 50°C$) is in between the glass transition of the veriflex ($T_g = 62°C$) and co-polyster ($T_g = 17°C$), which indicates the good compatibility between them.

3.4.2.3   Thermo Gravimetric Analysis

TGA measures the amount and rate of weight/mass change of a material with respect to temperature, time and atmosphere. TGA is used to study thermal and oxidative stability, composition (of multi-component polymer systems), decomposition kinetics and volatile or moisture content. TGA instrument consists of a precise balance and a programmable furnace and rely on the accurate measurement of mass change, temperature, and temperature change. An inert environment to prevent oxidation is maintained by purging the chamber with inert gas, such as nitrogen, helium or argon.

Yang et al. (2006) used TGA to quantify the water content to study the effect of moisture on the properties of the SMP. Thermal stability of high glass-transition thermoplastic SMP polyimides (PIs) and thermoset SMP PIs was found to be 510°C and 520°C, respectively, by Xiao et al. (2015). Du and Gan (2014) confirmed that the grafting polymerisation on the surface of hydroxyapatite (HAp) nanoparticles to the poly(D,L) (PDLLA) matrix was successfully performed with the help of TGA. They observed that the increase in HAp concentration in the reaction mixture decreased the initiation efficiency of the polymerisation. Further, TGA can also be used to find the concentration of filler/reinforcement.

3.4.3   Imaging-Based Characterisation

Electron microscopes are valuable tools that provide information about the structure, morphology, topology, composition and crystallinity in sub-micron level. They allow visualisation and analysis in the realms of micro-space to nanospace. The resolution of the light microscope is limited by the wavelength of light; however, electron microscope uses electrons as the source and their lower wavelength helps in resolving in greater detail. The user should consider desired resolution and information required for choosing the correct technique. For polymeric samples, a desirable image must balance image contrast, image quality and radiation damage.

SEM is used for the characterisation of polymer surfaces and surface topography. Internal structures can also be studied by fracturing the sample. The resolution of

SEM is comparatively less than TEM. However, SEM images have excellent depth of field, which allows it to produce accurate 3D representation of the sample's surface. Sample preparation for SEM is much simpler than TEM with specimen mostly mounted on aluminium holder with an electro-conductive adhesive. SEM techniques are widely used by the researchers to study the dispersion of reinforcement and filler in the polymer matrix. Ash et al. (2004) used FE–SEM images of the fractured surface to support their hypothesis that incorporation of alumina nanoparticles in the PMMA matrix decreased the glass-transition temperature of the polymer. They observed the de-bonding of the large alumina particles from the PMMA matrix following tensile test indicating the poor wetting of the nano-alumina and PMMA. They concluded that the voids increased the mobility of the bulk polymer leading to decreased overall T_g.

In transmission electron microscope, electron beams pass through the sample to provide information about the structure and the sample has to be very thin (less than 100 nm) or mounted on thin copper grids. Since, the polymers consist primarily of low atomic number elements, the contrast for imaging of polymer using TEM is difficult to obtain. Further, damage by radiation on polymers due to the high voltage required to accelerate the electrons is of concern (Libera and Egerton 2010). However, established techniques such as bright and dark-field imaging, electron diffraction, high-resolution imaging and analytical microscopy are used to study polymers. Contrast mechanism, resolution and specimen preparation for testing polymer in TEM has been comprehensively covered by Michler (2008) and Saywer et al. (2008). Composite SMP containing heavy elements is easier to image since the varying electron density leads to better contrast. Cao and Jana (2007) investigated dispersion and exfoliation of nanoclay-tethered shape-memory PU nanocomposite using TEM images and further validated the exfoliation of 1% nanoclay SMP composite using wide-angle X-ray scattering (WAXS). TEM was also used to study the phase separation in PU of hard segments and soft segments in the nanoscale. The sample was stained with 0.5% ruthenium oxide (RuO_4) for 2 h after fixing on the copper grid. The hard segments (urethane) are depicted as very small dark particles, whereas the soft segments are seen as bright matrix.

The microscopy methods provide important qualitative information about the samples. However, a volumetric quantification of structure is not possible. As a result, additional methods that provide quantitative information are required along with the microscopy data for better understanding.

3.4.4 STRUCTURAL AND CHEMICAL CHARACTERISATION

3.4.4.1 Scattering Techniques for Characterisation

Wide- and small-angle X-ray scattering (WAXS and SAXS) are suitable to study the crystalline domains involved in the SMP. They provide a quantitative description of crystalline domains in terms of crystal size, structure and orientation. WAXS provides information about the polymer chain segments, for example, crystalline size, structure, degree of crystallinity, crystal distortions and orientation of crystalline and amorphous space. WAXS, in general, can be used to interpret the influence of crystal structure and crystal formation on SME in semi-crystalline polymers.

However, its application is somewhat limited by the relative difficulty in translating data to a structural determination.

When an incident X-ray interacts with a material in the WAXS technique, the electrons begin to oscillate at the same frequency as the incoming beam. If the atoms have no regular arrangement (i.e. in entirely amorphous materials), the oscillation of the electrons will destructively interfere with one another, and little to no energy leaves the system. If there is a regular arrangement of atoms (i.e. crystalline domains), constructive interference results in X-ray beams leaving the sample in well-defined direction. When studying a polymeric sample, the amorphous regions produce very broad peaks in the diffraction spectrum and crystalline regions produce sharp peaks. Mass fraction of crystallinity of the sample can be calculated by comparing the area under the curves in these regions. Typically, nanostructures (e.g. domain sizes and long periods in semi-crystalline materials, thermoplastic elastomers) are observed through SAXS.

Liu et al. (2002) used WAXS to investigate the influence of cross-linking on polycyclooctene (PCO) micro-structure. They noted amorphous halo and four crystalline diffraction rings, which showed nearly constant d-spacing for the PCO samples even with different amount of peroxide cross-linking. The crystallinity was decreased with increasing the concentration of cross-linking compound due to constrained crystal growth. Further, it was observed that the rate and extent of strain recovery were increased with cross-linking density and the results helped to tailor the parameters for a given application, improving the flexibility of the system.

Ji et al. (2006) concluded that wet spinning of SMPU increased the crystallinity content compared to bulk material using WAXS characterisation. Further, SAXS was also used to conclude that concentration of hard segment micro-domains was increased and partial molecular orientation was observed in wet-spinned SMPU fibres than the bulk SMPU.

SAXS measurements were carried out on reverse bi-directional shape memory, poly(ω-pentadecalactone) PPD–PCL poly(ϵ-caprolactone) synthesised by Behl et al. (2013) to study the nanoscale level structural changes. They observed similar scattering patterns for shape A at T_{high} and shape B at T_{low} even after several cycles, which confirmed the extreme reversibility of reverse bi-directional SME of the PPD–PCL. They established that PPD phase remained unchanged during reverse bi-directional SME and attributed the change in the long period determined from the SAXS analysis of the two shapes to the semi-crystalline PCL phase.

The combination of SAXS and WAXS enables the simultaneous observation of two consecutive structural size levels and gives insights into structural modification during programming and recovery. By correlating, thermomechanical parameters (elongation and temperature), chemical parameters (network chain lengths) and the structural parameters (crystallite size, domain size and arrangement), the SME can be explained for different polymers from the nanometre scale to the micrometre scale.

3.4.4.2   Fourier Transform Infrared Spectroscopy

Infrared spectroscopy is a very important tool to investigate the polymer structures at the molecular scale. The bonding arrangement and chemical composition of constituents in a polymer, polymer blends and polymer composite can be obtained using infrared spectroscopy. FTIR is used for identification of various functional groups including covalently bonded groups.

When IR radiation, typically in the mid-IR range (4000–400 cm^{-1}) is passed through a polymer sample, it absorbs energy to vibrate or stretch different bonds corresponding to a chemical structure. These spectra act as a fingerprint for particular type of bonding, allowing for its identification. The spectrometer operates in a transmission mode or reflection mode or attenuated total reflectance (ATR) mode. The transmission is the most common and oldest mode and is excellent for solid, liquid and gases. However, the sample preparation can be difficult with KBr-pressed disc technique. The reflection mode is mostly used when polymer is not well dissolved at room temperature, and a pellet or film is characterised. In the ATR mode, the IR beam reflects from an interface via total internal reflection. The information provided by the ATR–FTIR is of the surface rather than the bulk material.

FTIR is a versatile tool for SMP applications. FTIR can provide information about the functional groups involved in shape-memory transitions. Further, it can track the polymerisation process of a polymer by monitoring the functional groups to be evolved or consumed in real time. FTIR can help in optimisation of the reactions by measuring the reaction as a function of temperature, time, concentrations and so on (Moraes et al. 2008).

Deka et al. (2010) employed FTIR to investigate the linkages between hyperbranched PU and MWCNT. It was concluded that the overlapping of –NH stretching of urethane and –OH stretching of modified MWCNT and shifting of –C=O band width increase in concentration of MWCNT confirmed the interaction between polymer and nanofillers.

The interaction of moisture with PU SMP was investigated by Yang et al. (2006) with the help of FTIR spectroscopy to study the mechanism affecting the glass-transition temperature of the polymer. It was observed that the hydrogen bonded N–H stretching shifts to higher frequency and the bonded C=O stretching shifts to lower frequency with increased intensity, when the sample immersion time in water is increased. After heating, the peaks returned to the original position of the dried SMP. It was concluded that the water absorption weakened the hydrogen bonding between N–H and C=O groups of the ester-based PU SMP, which caused to decrease the glass-transition temperature.

3.4.4.3 Nuclear Magnetic Resonance Technique

NMR technique is one of the most powerful methods for the analysis of organic compounds. NMR spectroscopy provides information about chemical composition, cross-link density, repeating units and distribution, branching, molecular weight and tacticity, degree of conversion and shape-memory transitions at molecular level. Bertmer et al. (2005) investigated the SME in a covalent cross-linked system obtained from UV cross-linking of poly [L-(lactide)-ran-glycolide)] dimethacrylates by double-quantum excitation method, which reflects the strength of dipolar coupling in the network. The temporary shape of SMP fixed by heating and elongation revealed 80% higher dipolar coupling compared to the permanent shape. In temporary shape, the segments between cross-linking points are stretched and partially aligned, resulting in greater dipole moment than the permanent shape.

Powers et al. (2008) investigated the structure and dynamics of carbon nanofibre (CNF)-reinforced PU composite using a high-resolution solid state H–NMR. The CNF led to shifting and broadening of the signals with increase in concentration of CNF, whereas the proton spin-lattice and spin–spin relaxation time were not significantly altered. The broadening was inhomogeneous and related to the difference in magnetic susceptibility between the thermoplastic elastomer and the CNF. Further, proton-spin diffusion revealed about the stress-induced crystallinity of the sample.

3.4.5   SHAPE-MEMORY EFFECT CHARACTERISATION

The SME at the macroscopic level is scrutinised by fixing of the secondary shape and recoverability of the permanent shape. These properties are affected by the thermal conditions such as deformation temperature, fixing temperature, mechanical deformation variables such as strain rate and strain. The SME is quantified by three parameters: (1) shape fixity ratio, (R_f), represents the ability of the polymer to retain the secondary/programmed shape; (2) shape-recovery ratio, (R_r), describes the ability of the material to recover its permanent shape; and (3) recovery stress – the stress generated by the polymer due to constrained recovery. These parameters are usually studied by thermomechanical experiments or bending test.

3.4.5.1   Thermomechanical Cyclic Test

The thermomechanical testing on shape-memory properties can be carried out on a universal tensile machine or a DMA (Xie 2010, Chen 2017), by following tailored test procedures. The test procedure consists of programming module, where secondary shape is given, and a recovery module, where the recovery to permanent shape is studied. The programming module can be carried out in stress-controlled or strain-controlled conditions, whereas the recovery module can be performed under unconstrained conditions, isostrain conditions or partially constrained conditions.

The protocol for programming modules consist of three steps: (1) Heating the polymer above the switching temperature, $T_{sw} < T_{prog}$ and then stretching the polymer to a pre-defined strain, ε_m at certain strain rate (strain-controlled test) or tensile stress σ_m is applied at a defined stress rate to get the extension (stress-controlled test) and held for a fixed period of time; (2) Cooling the sample below switching temperature of the sample, $T_c < (T_{sw}-15K)$ at a certain rate while keeping the ε_m (strain controlled) or σ_m (stress controlled) constant for fixing the secondary shape; (3) Release the holding force at T_c. The temperature must be held constant for a period of time, wherever there is a change in temperature to ensure consistent temperature in the sample. Subsequently, the polymer needs to be heated above the switching temperature under no constraints for recovering the original/permanent shape. This completes a thermomechanical cycle and these protocols can be run a number cycles to study the life cycle of the SMP.

A thermomechanical strain-controlled programming module with unconstrained recovery test module of SMPU performed on DMA is shown in Figure 3.11. The sample was heated to 80°C and elongated to 25% strain (ε_m) and then cooled down to −40°C for fixing the shape. The force holding the elongation is released and the strain recorded is taken as ε_u. Finally, the sample is again heated back to 80°C unconstrained and the unrecovered strain is ε_p.

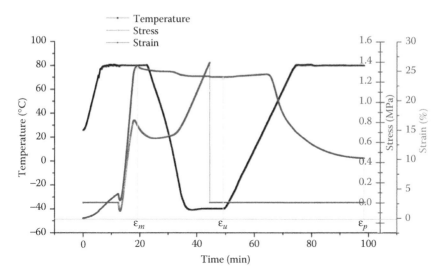

FIGURE 3.11 Temperature, stress, strain versus time graph for shape-memory test of SMPU.

Shape fixity ratio, R_f, is defined as the ratio between the strain held after programming the polymer (withdrawal of the deforming force) to the maximum strain applied.

Strain-controlled test

$$R_f(N) = \frac{\varepsilon_u(N)}{\varepsilon_m}$$

Stress-controlled test

$$R_f(N) = \frac{\varepsilon_u(N)}{\varepsilon_1(N)}$$

where:

$\varepsilon_u(N)$ is the strain held after cooling and withdrawal of the force for Nth cycle

ε_m is the maximum strain applied during programming

$\varepsilon_1(N)$ is the maximum strain at σ_m after cooling for the Nth cycle

Strain recovery is the ratio between the strain recovered after heating to the maximum strain applied.

Strain controlled

$$R_r(N) = \frac{\varepsilon_m - \varepsilon_p(N)}{\varepsilon_m - \varepsilon_p(N-1)}$$

Stress controlled

$$R_r(N) = \frac{\varepsilon_m - \varepsilon_p(N)}{\varepsilon_m - \varepsilon_p(N-1)}$$

ε_m is the maximum strain applied during programming
$\varepsilon_1(N)$ is the maximum strain at σ_m after cooling for the Nth cycle
$\varepsilon_p(N)$ is the irrecoverable strain after heating above switching temperature for Nth cycle

Recovery stress generated is quantified by heating the polymer under load constraint or isostrain conditions during recovery. Recovery stress is the stress applied by the polymer to return to its original shape. SMP have very low recovery stress compared to shape-memory alloys making it unusable for most applications. Huang et al. (2010) observed that a higher programming deformation/strain and using fillers such as carbon black resulted in a higher recovery stress and lesser recovery ratio of the polymers.

The influence of strain-holding conditions on shape recovery of PU shape memory was studied by Tobushi et al. (2006). They reported that if the SMP foam is held constrained above T_{sw} after programming module, new shape forming appears. The thermal motion of the molecular chains is active above T_{sw}, which causes re-orientation of the molecular chains decreasing the shape recovery of the SMP. The recovery depends on the holding strain, temperature and time. They also reported that holding time and maximum strain had no effect on the shape fixity and recovery if the temporary shape is kept below T_{sw}.

In real-life application, SMP has to recover under an applied stress. Lakhera et al. (2012) performed partially constrained recovery characterisation on meth(acrylate) SMP networks, where it recovered under constraining stress condition. They found that there was a linear relationship between the strain recovered and the constraining stress applied during recovery.

3.4.5.2 Bending Test

The bending test is a simple and easy analysis of SME of the polymers. Many researchers have used this simple technique to investigate the SME of their newly synthesised SMP.

In the bending test, a flat sample is heated above the T_{sw} and it is bent at an angle θ_0 (usually in a semicircle) and held in that shape. The schematic presentation of the same is shown in Figure 3.12. The deformed sample is cooled below the T_{sw} and the holding force is released. Finally, the sample is heated to above T_{sw} and the recovery of the polymer to the flat shape is recorded in terms of a series of deforming angles, θ against time. The shape recovery ratio, R_r is calculated from the angle after recovery, θ_N and the deformation angle, θ_0

$$R_r = \frac{(\theta_0 - \theta_N)}{\theta_0}$$

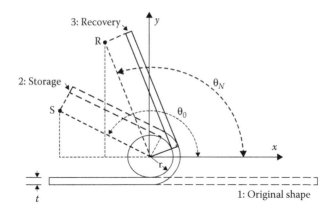

FIGURE 3.12 Bending test for shape-memory characterisation. (From Leng, J. et al., *Prog. Mater. Sci.*, 56, 1077–1135, 2011.)

Zhang and Ni (2007) conducted bending recoverability test of their developed SMP/carbon-fibres (CF) laminates. They observed that the recovery of SMP/CF fabric laminate was largest, whereas lowest in the CF/SMP laminates. This is attributed to the elastic stiffness of the SMP and carbon fabric. Since the resultant bending moment was in the recovery direction of SMP/CF laminate, the recovery was overshot and became more than 100%, whereas the CF/SMP laminate had smaller recovery rate. Further, they reported that self-weight of the laminate decreased the recovery of laminates. The advantage of bending test is easy sample preparation and testing procedure. However, very small strain is applied, that is, only minor plastic deformation is involved. Another issue is the non-uniform stress and strain that make it difficult to analyse (Wagermaier et al. 2010).

Many researchers have reported shape-recovery properties by testing in conditions similar to the end applications. Hasan et al. (2016b) studied the shape recovery of synthesised SMP nanocomposites foams for embolisation of cerebral aneurysms. Cylindrical foam samples were prepared with a diameter of 4 mm and a height of 10 mm and were radially crimped to their smallest possible diameter after heating the sample above T_{sw}. During recovery, the sample was kept in a water bath at 50°C and diameter was measured after fixed interval of time. They reported increased recovery of 82% ± 13% at 5 wt.% loadings for Al_2O_3-filled SMPC foams. The highest recovery of 96% ± 20% was observed for 3 wt.% tungsten-filled foams. However, the shape recovery was reduced with increased filler concentration of SiO_2, which was attributed to its low density and increased quantity of nanoparticle in the matrix leading to disruption in polymer–polymer interactions.

3.5 CONCLUSION

Despite the early discovery of SMP, a significant progress in this field had only been made in the past three decades. Researchers have designed and synthesised new SMP and its composites for diverse applications using different synthesising route and manufacturing process. This chapter provides an overview of different manufacturing processes and synthesising methods for producing SMP. Further, various techniques for characterisation of SMP have been discussed briefly. The characterisation of important material properties and the techniques best suited to investigate them are highly application-specific and that SMP validation should be a constant process of feedback and refinement, rather than a single event.

REFERENCES

Ahir S.V., Tajbakhsh A.R., and Terentjev E.M. (2006). Self-assembled shape-memory fibers of triblock liquid-crystal polymers. *Advanced Functional Materials* 16 (4), 556–560.

Alteheld A., Feng Y., Kelch S., and Lendlein A. (2005). Biodegradable, amorphous copoly-ester-urethane networks having shape-memory properties. *Angewandte Chemie International Edition* 44, 1188–1192.

Ash B.J., Siegel R.W., and Schadler L.S. (2004). Glass-transition temperature behavior of alumina/PMMA nanocomposites. *Journal of Polymer Science B: Polymer Physics* 42, 4371–4383.

Bao M., Lou X., Zhou Q., Dong W., Yuan H., and Zhang Y. (2014). Electrospun biomimetic fibrous scaffold from shape memory polymer of PDLLA-co-TMC for bone tissue engi-neering. *ACS Applied Material Interfaces* 6 (4), 2611–2621.

Bao M., Wang X., Yuan H., Lou X., Zhao Q., and Zhang Y. (2016). HAp incorporated ultra-fine polymeric fibers with shape memory effect for potential use in bone screw hole healing. *Journal of Material Chemistry B* 4, 5308–5320.

Behl M., Kratz K., Zotzmann J., Nöchel U., and Lendlein A. (2013). Reversible bidirectional shape-memory polymers. *Advanced Materials* 25, 4466–4469.

Bertmer M., Buda A., Blomenkamp-Höfges I., Kelch S., and Lendlein A. (2005). Biodegradable shape-memory polymer networks: Characterization with solid-state NMR. *Macromolecules* 38 (9), 3793–3799.

Bouchard J., Cayla A., Odent S., Lutz V., Devaux E., and Campagne C. (2016). Processing and characterization of polyethersulfone wet-spun nanocomposite fibres containing multi-walled carbon nanotubes. *Synthetic Metals* 217, 304–313.

Cao F., and Jana S.C. (2007). Nanoclay-tethered shape memory polyurethane nanocompos-ites. *Polymer* 48 (13), 3790–3800.

Cha D.I., Kim H.Y., Lee K.H., Jung Y.C., Cho J.W., and Chun B.C. (2005). Electrospun non-wovens of shape-memory polyurethane block copolymers. *Journal of Applied Polymer Science* 96 (2), 460–465.

Chen T. *Characterization of Shape-Memory Polymers by DMA*. TA Instruments, New Castle, DE. http://www.tainstruments.com/pdf/literature/TA374%20Characterization%20of%20 Shape-Memory%20Polymers%20by%20DMA.pdf (Accessed June 14, 2017).

Cui Y., Tan M., Zhu A., and Guo M. (2014). Mechanically strong and stretchable PEG-based supramolecular hydrogel with water-responsive shape-memory property. *Journal of Material Chemistry B* 2, 2978–2982.

Defize T., Riva R., Raquez J.M., Dubois P., Jérôme C., and Alexandre M. (2011). Thermoreversibly crosslinked poly(e-caprolactone) as recyclable shape-memory poly-mer network. *Macromolecular Rapid Communications* 32 (16), 1264–1269.

Deka H., Karak N., Kalita R.D., and Buragohain A.K. (2010). Biocompatible hyperbranched polyurethane/multi-walled carbon nanotube composites as shape memory materials. *Carbon* 48 (7), 2013–2022.

Du K., and Gan Z. (2014). Shape memory behaviour of HA-g-PDLLA nanocomposites pre-pared via in situ polymerization. *Journal of Materials Chemistry B* 2, 3340–3348.

Gong T., Li W., Chen H., Wang L., Shao S., and Zhou S. (2012). Remotely actuated shape mem-ory effect of electrospun composite nanofibers. *Acta Biomaterialia* 8 (3), 1248–1259.

Gupta B.S. (2013). Manufacture, types and properties of biotextiles for medical applications, in *Biotextiles as Medical Implants*. King M.W., Gupta B.S., and Guidoin R. (Eds.), Woodhead Publishing.

Hager M.D., Bode S., Weber C., and Schubert U.S. (2015). Shape memory polymers: Past, present and future developments. *Progress in Polymer Science* 49, 3–33.

Hampikian J.M., Heaton B.C., Tong F.C., Zhang Z., and Wong C.P. (2006). Mechanical and radiographic properties of a shape memory polymer composite for intracranial aneurysm coils. *Materials Science and Engineering: C* 26 (8), 1373–1379.

Hasan S.M., Harmon G., Zhou F., Raymond J.E., Gustafson T.P., Wilson T.S., and Maitland D.J. (2016a). Tungsten-loaded SMP foam nanocomposites with inherent radiopacity and tunable thermo-mechanical properties. *Polymers Advanced Technologies* 27, 195–203.

Hasan S.M., Thompson R.S., Emery H., Nathan A.L., Weems A.C., Zhou F., Monroe M.B.B., and Maitland D.J. (2016b). Modification of shape memory polymer foams using tungsten, aluminum oxide, and silicon dioxide nanoparticles. *RSC Advances* 6, 918–927.

Hu J., Meng Q., Zhu Y., Lu J., and Zhuo H. (2009). Shape memory fibers prepared via wet, reaction, dry, melt, and electro spinning. US 20090093606 A1.

Huang W.M., Yang B., Zhao Y., and Ding Z. (2010). Thermo-moisture responsive polyurethane shape-memory polymer and composites: A review. *Journal of Material Chemistry* 20 (17), 3367–3381.

Ji F.L., Zhu Y., Hu J.L., Liu Y., Yeung L.Y., and Ye G.D. (2006). Smart polymer fibers with shape memory effect. *Smart Materials and Structures* 15 (6), 1547–1554.

Korley L.T.J., Pate B.D., Thomas E.L., and Hammond P.T. (2006). Effect of the degree of soft and hard segment ordering on the morphology and mechanical behavior of semicrystalline segmented polyurethanes. *Polymer* 47, 3073–3082.

Kunzelman J., Chung T., Mather P.T., and Weder C. (2008). Shape memory polymers with built-in threshold temperature sensors. *Journal of Material Chemistry* 18 (10), 1082–1086.

Lakhera N., Yakacki C.M., Nguyen Thao D., and Frick C.P. (2012). Partially constrained recovery of (meth)acrylate shape-memory polymer networks. *Journal of Applied Polymer Science* 126, 72–82.

Landsman T.L., Touchet T., Hasan S.M., Smith C., Russell B., Rivera J., Maitland D.J., and Cosgriff-Hernandez E. (2017). A shape memory foam composite with enhanced fluid uptake and bactericidal properties as a hemostatic agent. *Acta Biomaterialia* 47, 91–99.

Lawrence C. (2015). *Fibre to Yarn: Filament Yarn Spinning in Textile and Fashion*, Sinclair R. (Ed.), Woodhead Publishing, pp. 213–253.

Lendlein A., and Langer R. (2002). Biodegradable, elastic shape-memory polymers for potential biomedical applications. *Science* 296 (5573), 1673–1676.

Leng J., Lan X., Liu Y., and Du S. (2011). Shape-memory polymers and their composites: Stimulus methods and applications. *Progress in Materials Science* 56 (7), 1077–1135.

Li X., Hao X., Na H. (2007). Preparation of nanosilver particles into sulfonated poly(ether ether ketone) (SPEEK) nanostructures by electrospinning. *Materials Letters* 61 (2), 421–426.

Libera M.R., and Egerton R.F. (2010). Advances in the transmission electron microscopy of polymers. *Polymer Reviews* 50 (3), 321–339.

Litvinov V.M., Xu J., Melian C., Demco D.E., Möller M., and Simmelink J. (2011). Morphology, chain dynamics, and domain sizes in highly drawn gel-spun ultrahigh molecular weight polyethylene fibers at the final stages of drawing by SAXS, WAXS, and 1H solid-state NMR. *Macromolecules* 44 (23), 9254–9266.

Liu C., Chun S.B., Mather P.T., Zheng L., Haley E.H., and Coughlin E.B. (2002). Chemically cross-linked polycyclooctene: Synthesis, characterization, and shape memory behavior. *Macromolecules* 35 (27), 9868–9874.

Liu C., Qin H., and Mather P.T. (2007). Review of progress in shape-memory polymers. *Journal of Material Chemistry* 17, 1543–1558.

Lu H., Liang F., and Gou J. (2011a). Nanopaper enabled shape-memory nanocomposite with vertically aligned nickel nanostrand: Controlled synthesis and electrical actuation. *Soft Matter* 7 (16), 7416–7423.

Lu H., Liang F., Yao Y., Gou J., and Hui D. (2014b). Self-assembled multi-layered carbon nanofiber nanopaper for significantly improving electrical actuation of shape memory polymer nanocomposite. *Composites Part B: Engineering* 59, 191–195.

Lu H., Liu Y., Gou J., Leng J., and Du S. (2010). Electrical properties and shape-memory behavior of self-assembled carbon nanofiber nanopaper incorporated with shape-memory polymer. *Smart Materials and Structures* 19 (19), 75021–75027.

Lu H., Liu Y., Gou J., Leng J., and Du S. (2011b). Surface coating of multi-walled carbon nanotube nanopaper on shape-memory polymer for multifunctionalization. *Composites Science Technology* 71 (11), 1427–1434.

Lu H., Yao Y., Huang W.M., Leng J., and Hui D. (2014a). Significantly improving infra-red light-induced shape recovery behavior of shape memory polymeric nanocomposite via a synergistic effect of carbon nanotube and boron nitride. *Composites Part B: Engineering* 62, 256–261.

Lu X., Chan C.Y., Lee K.I., Ng P.K., Fei B., Xin J.H., and Fu J. (2014c). Super-tough and thermo-healable hydrogel – Promising for shape-memory absorbent fiber. *Journal of Material Chemsitry B* 2 (43), 7631–7638.

Makowski R.L., and Lundberg H.S. (1990). Plasticization of metal sulfonate-containing EPDM with stearic acid derivatives. *Ions in Polymers: Advances in Chemistry Series* 187 (1), 37–51.

Maksimkin A., Kaloshkin S., Zadorozhnyy M., and Tcherdyntsev V. (2014). Comparison of shape memory effect in UHMWPE for bulk and fiber state. *Journal of Alloys and Compounds* 586 (1), S214–S217.

Mendez J., Annamalai P.K., Eichhorn S.J., Rusli R., Rowan S.J., Foster J.E., Weder C. (2011). Bioinspired mechanically adaptive polymer nanocomposites with water-activated shape-memory effect. *Macromolecules* 44, 6827–6835.

Meng Q., and Hu J. (2008). Self-organizing alignment of carbon nanotube in shape memory segmented fiber prepared by in situ polymerization and melt spinning. *Composites: Part A* 39, 314–321.

Meng Q., Hu J., and Zhu Y. (2007b). Shape-memory polyurethane/multiwalled carbon nanotube fibers. *Journal of Applied Polymer Science* 106, 837–848.

Meng Q., Hu J., Zhu Y., Lu J., and Liu Y. (2007a). Morphology, phase separation, thermal and mechanical property differences of shape memory fibres prepared by different spinning methods. *Smart Materials and Structures* 16 (4), 1192–1197.

Michler G.H. (2008). *Electron Microscopy of Polymers*. Springer-Verlag, Leipzig, Germany.

Moraes L.G.P., Rocha R.S.F., Menegazzo L.M., Araújo E.B., Yukimito K., and Moraes J.C.S. (2008). Infrared spectroscopy: a tool for determination of the degree of conversion in dental composites. *Journal of Applied Oral Science* 16 (2), 145–149.

Ni Q.Q., Zhang C., Fu Y., Dai G., and Kimura T. (2007). Shape memory effect and mechanical properties of carbon nanotube/shape memory polymer nanocomposites. *Composite Structures* 81 (2), 176–184.

Nji J., and Li G. (2010a). A biomimic shape memory polymer based self-healing particulate composite. *Polymer* 51 (25), 6021–6029.

Nji J., and Li G. (2010b). A self-healing 3D woven fabric reinforced shape memory polymer composite for impact mitigation. *Smart Materials and Structures* 19 (3), 1–7.

Ohki T., Ni Q., Ohsako N., and Iwamoto M. (2004). Mechanical and shape memory behavior of composites with shape memory polymer. *Composites Part A: Applied Science and Manufacturing* 35 (9), 1065–1073.

Oyama H.T. (2009). Super-tough poly(lactic acid) materials: Reactive blending with ethylene copolymer. *Polymer* 50 (3), 747–751.

Powers D.S., Vaia R.A., Koerner H., Serres J., and Mirau P.A. (2008). NMR characterization of low hard segment thermoplastic polyurethane/carbon nanofiber composites. *Macromolecules* 41 (12), 4290–4295.

Richardson T., Mosiewicki A.M., Uzunpinar C., Aranguren M.I., Kilinc-Balci F., Broughton R.M., and Auad M.L. (2011). Study of nanoreinforced shape memory polymers processed by casting and extrusion. *Polymer Composites* 32, 455–463.

Saywer L.C., Grubb D.T., and Meyers G.F. (2008). *Polymer Microscopy*, 3rd ed. Springer, New York.

Tobushi H., Hayashi S., Hoshio K., and Miwa N. (2006). Influence of strain-holding conditions on shape recovery and secondary-shape forming in polyurethane-shape memory polymer. *Smart Materials and Structures* 15 (4), 1033–1038.

Tseng L.F., Mather P.T., and Henderson J.H. (2013). Shape-memory-actuated change in scaffold fiber alignment directs stem cell morphology. *Acta Biomaterialia* 9 (11), 8790–8801.

Unger R.E., Peters K., Huang Q., Funk A., Paul D., and Kirkpatrick C.J. (2005). Vascularization and gene regulation of human endothelial cells growing on porous polyethersulfone (PES) hollow fiber membranes. *Biomaterials* 26, 3461–3469.

Wache H.M., Tartakowska D.J., Hentrich A., and Wagner M.H. (2003). Development of a polymer stent with shape memory effect as a drug delivery system. *Journal of Material Science: Materials in Medicine* 14, 109–111.

Wagermaier W., Kratz K., Heuchel M., and Lendlein A. (2010). Characterization methods for shape-memory polymers. *Advances in Polymer Science* 226, 97–145.

Weiss R.A., Izzo E., and Mandelbaum S. (2008). New design of shape memory polymers: Mixtures of an elastomeric ionomer and low molar mass fatty acids and their salts. *Macromolecules* 41 (9), 2978–2980.

Xiao X., Kong D., Qiu X., Zhang W., Zhang F., Liu L., Liu Y., Zhang S., Hu Y., and Leng J. (2015). Shape-memory polymers with adjustable high glass transition temperatures. *Macromolecules* 48 (11), 3582–3589.

Xie T. (2010). Tunable polymer multi-shape memory effect. *Nature* 464, 267–270.

Xie T. (2011). Recent advances in polymer shape memory. *Polymer* 52 (22), 4985–5000.

Xu H., An M., Lv Y., Zhang L., and Wang Z. (2016). Structural development of gel-spinning UHMWPE fibers through industrial hot-drawing process analyzed by small/wide-angle X-ray scattering. *Polymeric Bulletin* 1–16.

Yang B., Huang W.M., Li C., and Chor J.H. (2005). Effects of moisture on the glass transition temperature of polyurethane shape memory polymer filled with nano-carbon powder. *European Polymer Journal* 41 (5), 1123–1128.

Yang B., Huang W.M., Li C., and Li L. (2006). Effects of moisture on the thermomechanical properties of a polyurethane shape memory polymer. *Polymer* 47 (4), 1348–1356.

Zhang C., and Nib Q. (2007). Bending behavior of shape memory polymer based laminates. *Composite Structures* 78 (2), 153–161.

Zhu G., Liang G., Xu Q., and Yu Q. (2003). Shape-memory effects of radiation crosslinked poly(ε-caprolactone). *Journal of Applied Polymer Science* 90, 1589–1595.

Zhu Y., Hu J., Yeung L.Y., Liu Y., Ji F., and Yeung K. (2006). Development of shape memory polyurethane fiber with complete shape recoverability. *Smart Materials and Structures* 15 (5), 1385–1394.

Zhu Y., Hu J., Yeung L.Y., Lu J., Meng Q., Chen S., and Yeung K. (2007). Effect of steaming on shape memory polyurethane fibers with various hard segment contents. *Smart Materials and Structures* 16 (4), 969–981.

Zhuo H., Hu J., and Chen S. (2008). Preparation of polyurethane nanofibers by electrospinning. *Journal of Applied Polymer Science* 109, 406–411.

4 Microwave Processing of Polymer Matrix Composites

S. Zafar and A.K. Sharma

CONTENTS

4.1 INTRODUCTION

Today there is a need for optimum utilisation of the natural resources to minimise the threat to the ecological system together with other environment-related concerns. The fast pace industrialisation has led to large-scale exploitation of natural resources. Metals have been an integral part of the human life in various applications such as construction, transport industries and defence applications. The replacement of metals would be a distant dream, but rapid progress in development of composites has proved to be a turning point. Composites are multi-phased systems that consist of at least two different groups of materials, which are chemically and physically distinct and separated by interfaces (Mantia and Morreale 2011). The socio-economic impact of composites has been so immense that it has become an area of active research and development, as no sector has been untouched from composites; be it household items toys, sports equipment, construction industry and aircraft structures (Chawla 2012). Composite materials have been regarded as a giant step in the ever-constant endeavour of optimisation of materials.

Since the early 1960s, there has been an increasing demand of lightweight, stiff and strong materials for fields as diverse as aerospace, energy and civil construction. The demands made on materials better for overall performance are so great and diverse that no materials can satisfy them. This has naturally led to the resurgence of the ancient concept of combing different materials in an integral-composite to satisfy the user requirements. The integration of the materials science and engineering with manufacturing and design has led from the conception of commissioning of an item, through inspection during lifetime, as well as failure analysis. Many processing techniques are available for manufacturing polymer composites. Curing or cross-linking occurs in thermosets by appropriate chemical agents by application of heat and pressure. In conventional processing, thermal energy is provided for this purpose. But there are several problems associated with the conventional processing techniques, such as thermal gradients, residual stresses and longer curing times. Residual stress can cause warping of laminates, fibres waviness, matrix micro-cracking and ply delamination. High-performance polymeric composites, reinforced with carbon, glass or aramid fibres have been effectively used in aerospace and electronics industries in applications requiring lightweight, high-specific strength stiffness, corrosion and chemical resistance and tailorable thermal expansion coefficient. The dielectric properties of glass or high-performance, polymeric-reinforced composites have made them attractive for printed circuit boards used in aviation, marine and land-based systems. Generic applications of polymeric composites have been hindered by their high cost of orientation (lay-up) and forming (moulding and curing). Innovative processing, including automated lamination, rapid consolidation and curing and out-of-autoclave processing, is being attempted to reduce the costs associated with processing. Microwave processing offers promising rapid, non-autoclave processing composite structures. Processing of very thick cross sections using conventional processing requires complex cure schedules with very slow thermal ramp rates and isothermal holds to control overheating due to cure reaction exotherms and poor thermal conductivity. Microwave radiation is penetrating and rapid, even heating characteristics, thick composites were initially targeted as ideal application of microwave processing. The mechanical properties were found to be at least equivalent to those conventionally processed materials, with property enhancement such as reduced void content occurred. Increased adhesion and improved mechanical properties at the fibre/matrix were observed in carbon-fibre (CF) composites due to preferential heating at the conductive fibres surface (National Research Council 1994).

Further, extensive use of composite materials in aerospace, automobiles, engineering structures, sports goods and so on have necessitated the need of efficient, time-saving and environment-friendly processes that can overcome the limitations associated with conventional processing techniques. Microwave material processing offers significant time saving and hence provides cost-effective solution, clean and environment-friendly solution. Microwave processing involves use of microwave radiations that offer a new approach to composite processing. In microwave processing the energy is directly injected in the material without reliance on conventional modes of heat transfer.

4.2   INTRODUCTION TO MICROWAVES

Microwaves are electromagnetic radiations in the frequency range of 0.3–300 GHz and wavelength of 1 m–1 mm as illustrated in Figure 4.1. Microwaves are coherent and polarised and can be transmitted, absorbed or reflected depending on the material type being processed (Clark and Sutton 1996). Originally, microwaves were used for telecommunication before, radar detection, tracking, electronic warfare, medical

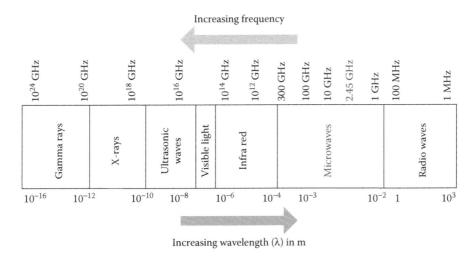

FIGURE 4.1 Electromagnetic spectrum showing various frequencies and wavelength.

treatment, non-destructive testing, power transmission, heating and so on. However, application of material heating through microwaves was first discovered in 1946 by Percy Spencer. It was later discovered that in microwave heating, energy conversion takes place instead of energy transfer through conventional modes of heat transfer. Since then microwaves have been extensively used in material processing due its attributes such as volumetric heating, reduced processing time, low power consumption and less environmental hazards (Jones et al. 2002). Until the year 2000, basically ceramics, ceramic composites, polymers, polymer composite and semi-conductors were processed with microwaves because they are good absorber of microwaves at room temperature (Aravindan and Krishnamurthy 1999; Ramkumar et al. 2002; Fang et al. 2008; Zhu et al. 2008; Budinger 2008). The frequencies reserved by the Federal Communication Commission (FCC) for heating in industrial, scientific and medical (ISM) systems are 915 and 2.45 GHz (Thostenson and Chou 1999).

In conventional processing, the heat is transferred through convection, conduction and radiation by virtue of thermal gradients between the source and the target. In contrast, microwave energy is delivered directly to materials through molecular interaction with the electromagnetic field. In microwave heating the transfer of heat is due to energy conversion, rather than energy transfer. This difference in the way how energy is delivered results in the potential benefits of using microwaves for material processing. In microwave heating, microwave energy penetrates into the materials and deposits energy. This deposited energy leads to generated heat throughout the volume of the material. As the transfer of energy does not rely on diffusion of heat from the surfaces, it is possible to achieve rapid and uniform heating of thick materials. In traditional heating, curing (cycle) time is often longer due to slow heating rates that are chosen to minimise steep thermal gradients. For polymers, which inherently have low thermal conductivities, microwave processing can significantly reduce processing times. As microwaves can transfer energy throughout the volume of the material, the processing time as well as overall product quality is enhanced.

4.3　THEORY OF MICROWAVE HEATING

A material can be heated in microwave radiation due to the volumetric, rapid and uniform heating. Microwaves consist of electric and magnetic field orthogonal to each other. The dominant mechanism of microwave-induced heating at 2.45 GHz involves the agitation of molecular dipoles due to the presence of an oscillating electric field (for non-magnetic materials). In the presence of an oscillating field, molecular dipoles re-orient themselves in order to be in phase with the alternating field. These orientations are restricted by molecular interaction forces, which increase the molecular kinetic energy. As kinetic energy increases, system temperature increases within a short time. This time period depends on the electrical, magnetic and physical properties of the heated material (Farag et al. 2012).

The bulk metallic materials tend to reflect microwaves at 2.45 GHz when processed at room temperature owing to their low *skin depth* (d_s). Skin depth in microwave material processing is defined as the distance from the surface at which the magnitude of the field strength drops by a factor of 1/e (36.8%) (Mondal et al. 2010). The skin depth is calculated using Equation 4.1:

$$d_s = \frac{1}{\sqrt{\pi \mu f \sigma}} = 0.029(\rho \lambda_0)^{1/2} \tag{4.1}$$

where:
σ is the electrical conductivity (S/m)
ρ is the electrical resistivity (Ω m)
λ_0 is the incident wavelength (mm)

Hence, it is imperative that absorption of microwave power depends on the dielectric properties of the materials. The microwave power absorbed per unit volume of the material exposed can be estimated using Equation 4.2 (Leonelli et al. 2008):

$$P = \omega \left(\varepsilon_0 \varepsilon'' E_{rms}^2 + \mu_0 \mu'' H_{rms}^2 \right) \tag{4.2}$$

where:
μ_0 is magnetic permeability of air (H/m)
μ'' is the imaginary component of the magnetic permeability (H/m)
E_{rms} is the root mean square of the electric field (V/m)
H_{rms} is the root mean square of the magnetic field (A/m)

In case of non-magnetic materials, μ'' is negligible, and hence contribution of magnetic field in power absorption can be neglected and Equation 4.2 reduces to a single term, $P = \omega(\varepsilon_0 \varepsilon'' E_{rms}^2)$. In magnetic material, the factors that contribute to heating include magnetic losses such as hysteresis, eddy current, domain wall and electron spin resonance. During microwave exposure, the electromagnetic energy penetrates into the target material. Ideally, volumetric heating should take place in the target material and the amount of energy added to the target material is given by Equation 4.3 (Roger 1998):

$$E = \int p.dT \quad \text{from 0 to } T \tag{4.3}$$

where:
  E is the total energy added to the target (J)
  p is the applied power (W)
  T is the time (s) for which total power is applied

Microwave hybrid heating (MHH) technique is utilised for materials that are poor absorbers of microwave radiation at low temperature. In this technique, a susceptor material is used to initiate heating. The susceptor couples with microwaves at room temperature and gets heated rapidly, which raises the temperature of the poor microwave-absorbing materials. The poor absorbing materials are heated beyond their critical temperature (T_c) through conventional modes of heat transfer, such as conduction, convection and radiation.

Microwave energy is transferred to materials through interaction of electromagnetic field at molecular level and ultimately the dielectric properties of materials determine the effect of the applied electromagnetic field on the material. The propagation of electromagnetic field in the material is governed by complex dielectric constant (ε^*) and is defined in Equation 4.4:

$$\varepsilon^* = \varepsilon' - j\varepsilon'' = \varepsilon_0(\varepsilon_r - j\varepsilon''_{eff}) \tag{4.4}$$

where:
  ε^* is the complex dielectric constant, which is a measure of ability of a dielectric
      to absorb or store the microwave energy
  ε' is the dielectric constant, characterising the penetration of microwave radiation
      in the target material
  ε'' is the dielectric loss factor that indicates the materials ability to store incident
      energy
  ε_0 is the permittivity of free space (8.86×10^{-12} F/m)
  ε_r is the relative dielectric constant, which is measure of the polarisability of the
      material
  ε''_{eff} is the effective dielectric loss vector
  j is complex imaginary unit and is equal to $(-1)^{1/2}$

Another important term for expressing the dielectric response of the target material is the *loss tangent*, and is defined in Equation 4.5.

$$\tan\delta = \frac{\varepsilon''}{\varepsilon'} \tag{4.5}$$

where, $\tan\delta$ indicates the ability of the material to convert the absorbed energy into heat energy.

In some materials, the magnetic dipoles are also able to couple with the magnetic component of the electromagnetic field and provide an additional heating mechanism. Thus, similar to the dielectric properties, the magnetic permeability, μ' and magnetic loss, μ'' can also be considered. The absorbed microwave power is converted into heat depending on electric field intensity, frequency, loss factor and permittivity. A lossy material gets heated more effectively than a low-loss material and rise in temperature (ΔT) with change in time (Δt) within the material is given by Equation 4.6 (Leonelli et al. 2008):

$$\frac{\Delta T}{\Delta t} = \frac{2\pi f \varepsilon''_{\text{eff}} |\mathbf{E}|^2}{\rho C_p} \tag{4.6}$$

Therefore, it is evident that higher the value of tanδ and ε'_r, smaller will be the depth of penetration. Large value of dielectric properties and higher frequency causes heating of the surface. While small value of dielectric properties and lower frequency will lead to volumetric heating of the sample. Materials having high conductivity and high permeability tend to have lower penetration depth. Therefore, the ability of dielectric material to absorb microwave energy changes during microwave processing. For example, the silicon carbide has a loss factor (ε'') of 1.71 at 2.45 GHz at room temperature, whereas for the same frequency, it is 27.99 at 695°C temperature (Metaxas 1991; Thostenson and Chou 1999).

4.4   MICROWAVE–MATERIAL INTERACTION

In microwave processing, absorption of microwaves by materials is volumetric and converts the energy into heat by friction heating by two modes: polarisation and ionic conduction. Polarisation is basically short-range displacement of charge, which generally occurs at high frequency (2.45 GHz) as illustrated in Figure 4.2. The dipoles present in the materials, rotate with alternating field. Ionic conduction involves long-range transport of charge, which encompasses oscillation ions that takes place at 915 MHz as shown in Figure 4.3. In composite material, the fundamental factor for heating mechanism is the Maxwell Wagner polarisation. Maxwell Wagner polarisation results from the polarisation accumulation of charges at the interface (Sutton 1989).

In addition to volumetric heating, microwave heating is also known for its selective heating characteristics. The molecular structure of the materials also affects the ability of microwaves to selectively interact with the materials and transfers energy. During exposure, microwaves will selectively couple with the higher loss material. This characteristic can be exploited to process multi-phased materials such as polymer composites. In multi-phased materials, some phases couple with microwaves more readily than the other. Direct microwave heating may offer advantages compared to the conventional processing, however, because of non-uniformity within the electric field may result in non-uniform heating. As the materials get processed under

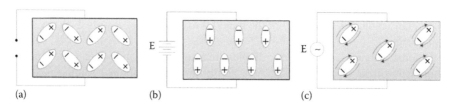

(a)   (b)   (c)

FIGURE 4.2   Redistribution of dipoles under (a) no electric field, (b) direct current field and (c) alternating current field.

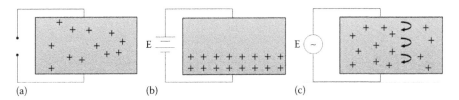

FIGURE 4.3 Redistribution of ionic charge under (a) no electric field, (b) direct current field and (c) alternating current field.

the microwave field, the changing dielectric properties lead to varying abilities of microwaves to generate heat. Sharp transformations in the ability of microwaves may cause difficulties in process modelling and control. In general, microwaves interact with different materials in different ways. Accordingly, materials can be classified by four different types with respect to their interaction with microwaves. Interaction of microwaves with different materials is illustrated in Figure 4.4.

Different materials exhibit different microwave absorption characteristics during interaction with microwaves due to the variation in the electric and magnetic field. The relationship between the dielectric loss factors and the ability to absorb microwave power per unit volume at room temperature for some materials are given in the Figure 4.5 (Ahmed and Siores 2001; Ku et al. 2002). Materials having low capacitance and high conductance (such as metals) have high dielectric loss factors. As the dielectric loss factors increases, the penetration depth approaches to zero. Materials

Interaction schematic	Material behaviour	Material characteristics	Examples	Applications
	Transparent	Low loss insulator	Teflon, glass, alumina	For making microwave fixtures
	Opaque	Conductor	Copper, aluminium, steel	Used in radar detection
	Absorber	Lossy insulator	Silicon carbide, charcoal, water	Used in microwave hybrid heating (MHH)
	Mixed absorber	Low loss insulator and absorbing material	Composite material (carbon fibre-reinforced composite)	Used in selective heating of materials

FIGURE 4.4 Behaviour of different materials towards microwave radiation.

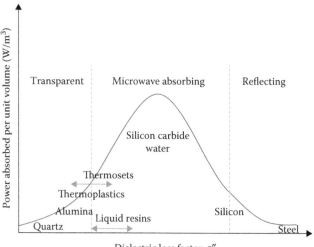

FIGURE 4.5 Microwave power absorption characteristics for some common materials. (From Thostenson, E.T., and T.-W. Chou, *Compos. A: Appl. Sci. Manuf.*, 30, 1055–1071, 1999.)

exhibiting such characteristics are known as reflectors. Materials having low dielectric loss factors exhibit very large penetration depth. Such materials are transparent to microwaves. Thus, microwaves interact with materials effectively that have dielectric loss factors in the middle of the conductivity range. In contrast, conventional heating occurs most efficiently with materials having high thermal conductivity. The characteristics, however, are not valid only for powdered material or bulk pre-treated metal. It has been observed that in powder form and at high temperature, all metals, ceramics and intermetallics start absorbing microwaves (Roy et al. 1999; Agrawal 2010).

4.5 HEAT TRANSFER MODES IN MICROWAVE HEATING

There are basically three modes of heat transfer in microwave heating, namely (1) direct heating, (2) selective heating and (3) hybrid heating. Direct heating mode is used to process materials that readily absorb microwave radiation (Figure 4.6a). The inherent inverse temperature gradient during microwaves processing causes development of formation of hotspots and non-uniform heating. Thermal instabilities lead to development of non-uniform properties and cracking in the material being processed. These materials absorb microwave energy at room temperature and convert it into heat through the mechanism of dipole rotation and ionic conduction. Examples of these materials are water, charcoal and silicon carbide. But problem arises while processing materials with microwaves at room temperature having lower depth of penetration. The second problem is that sometimes the temperature rise in an uncontrolled manner, which can lead to the phenomenon of thermal runway (Clark et al. 2000). This occurs mostly in ceramics such as Al_2O_3, SiO_2 and Fe_3O_4. The selective heating mode of microwave heating is accomplished through certain constraints. Special tooling or fixture is needed in order to achieve selective heating. For example, in microwave joining of bulk materials, the bulk materials are

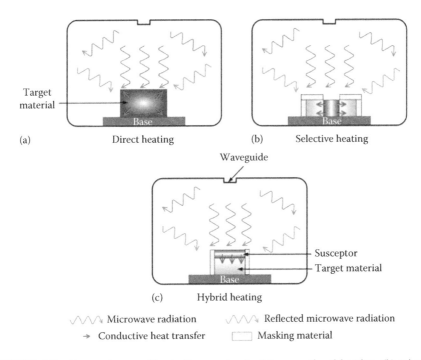

(a)   Direct heating   (b)   Selective heating

(c)   Hybrid heating

FIGURE 4.6 Temperature profiles in the samples for (a) conventional heating, (b) microwave heating and (c) microwave hybrid heating. (From Oghbaei, M., and O. Mirzaee, *J. Alloys Compd.*, 494, 175–189, 2010.)

masked using a masking material. The interfacing powder layer is exposed to microwave radiation. Thus, it is the powder that interacts with the microwave radiation, gets heated and consequently melted. The advantage of this technique is that only the desired part of the material is heated without heating the rest of the volume of the material (Figure 4.6b). The MHH heating mode is used to heat microwave-reflecting materials such as metallic materials. In hybrid heating, a special arrangement of a susceptor consists of a microwave susceptor. The susceptor is a microwave absorbing material, which couples with the incident microwave radiation at room temperature and gets heated up. The heated susceptor then transfers heat to the target material through conventional modes of heat transfer. Once the temperature of the target material goes beyond its critical temperature, it directly couples with the microwave radiation; heat is rapidly generated inside the core of the target material and flows from core to the surface (Figure 4.6c). Thus, in hybrid heating, a combined action of microwave coupled with heating of the susceptor material realise heating both from inside to outside and outside to inside, respectively. In conventional heating, the core is at lower temperature than outer surface and in microwave heating the core is at higher temperature than the outer surface. The former mode of heating results in poor micro-structural characteristics of the core (mainly powder compact in sintering process), whereas the later results in poor micro-structural characteristics of the surface. In hybrid heating, a relatively uniform temperature is obtained inside the

body due to heat supplied to the body at low temperature by the additional susceptor and at high temperature, the material itself starts absorbing microwave due to
increased dielectric loss. Consequently, a relatively uniform temperature inside the
body resulted. The hybrid heating is also used for processing of metallic materials, which reflect microwave at room temperature. The hybrid heating technique
is extensively used for materials having low dielectric loss at low temperature and
high dielectric loss at high temperature. This technique can be effectively applied for
processing of silicon nitride (Si_3N_4) and alumina (Al_2O_3). These ceramics are poor
absorbers of microwave up to a critical temperature; once the critical temperature is
reached, the material begins to couple with microwave as their dielectric properties
change. It is evident that absorption curve for both ionic conduction and dipole rotation, which mainly contributes to dielectric losses gets shifted to higher frequency
as temperature is increased (Figure 4.7). The metallic materials such as aluminium,

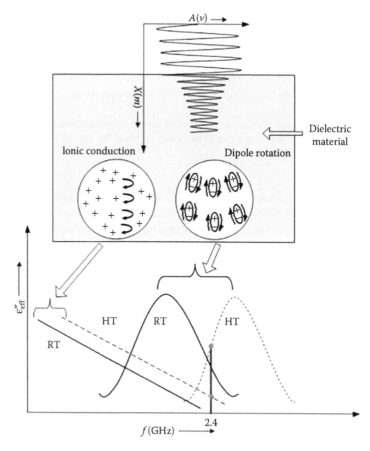

FIGURE 4.7   Ionic conduction and dipole rotation variation with temperature; x = distance;
A = amplitude of the electric field; ε''_{eff} = effective dielectric loss; f = frequency; RT = room
temperature; HT = high temperature. (Adapted from Clark, D.E. et al., *Mater. Sci. Eng. A*,
287, 153–158, 2000.)

copper and stainless steel are melted twice as fast as compared to conventional melting while using MHH (Chandrasekaran et al. 2011).

4.6 PHYSICS OF CURING POLYMER COMPOSITES USING MICROWAVE ENERGY

Microwave processing has been studied as an alternative processing method for composites due to beneficial effects on processing time, as well as the mechanical and thermal properties. Application of microwave processing has been difficult to cure relatively large samples. In contrast to conventional heating, microwave heating is volumetric in nature and is not restricted to the exposed surface area exposed. As a result, polymer matrix composite (PMC) can be processed more quickly using microwave processing route (Sgriccia and Hawley 2007). The principle mechanism of microwave absorption in polymers is the re-orientation of dipoles in the imposed electric field. The efficiency of microwave coupling with polymer materials depends on the dipole strength, mobility, mass and matrix state of the dipole. The dielectric constant can vary during processing cycle or if a phase change occurs as temperatures varies and the reaction proceeds changing the type and concentration of dipoles.

Polymer composites are the mixed microwave absorbers, where absorption of microwave energy is dependent on the dielectric properties of the matrix as well as reinforcements. Microwave couples with the phase that has the higher dielectric loss factor. The phase with high dielectric loss absorbs microwave energy, generates heat and heats the other consistent by conventional modes of heat transfer. The localised and selective heating in composite materials during microwave exposure improves the properties of the composites due to uniform heating than conventional methods. The concentration (volume percentage) of dielectric reinforcements in the matrix materials controls microwave absorption as well as dielectric constant, dielectric loss factor, loss tangent, relative permittivity and penetration. Significant changes in microwave absorption properties of PMCs has been reported, which is attributed to formation of conductive networks after a certain volume concentration due to perco-lation threshold (Mishra and Sharma 2016). The optimum concentration of dielectric additives provides control over the heating cycle of the polymer composites.

In case of PMC, for effective processing the dipole structure, frequency process-ing, temperature and characteristic properties of the reinforcements as well as of the matrix play an important role. In case of low absorbing reinforcements such as glass fibres (GFs), aramid, natural fibres, the dielectric properties of matrix dominate the heating due to microwave exposure. In case of fibres with high dielectric loss, such as CF, the fibre gets heated due to microwave and transfers heat to the polymer matrix by conduction. The polymer matrix absorbs microwaves during curing of fibre reinforcement whose dielectric loss factor is low. The dielectric properties of the matrix materials are affected by the change in the chemical structure during microwave exposure. For example, in thermosets, there is change in the viscosity during microwave exposure. The liquid matrix initially absorbs microwaves at room temperature; however, as the matrix begins to solidify, increasing viscosity prevents the dipole orientation. Thermosets also undergo cross-linking, which changes the internal

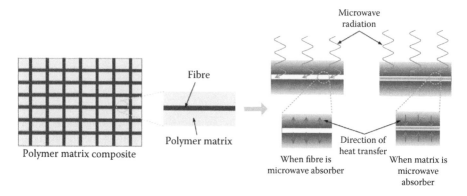

FIGURE 4.8 Mechanism of microwave curing of polymer composites.

network structure (Mishra and Sharma 2016). This reduces the microwave absorption in thermosets due to higher rate of cross-linking. Permittivity and dielectric loss factor of thermosets generally increases with temperature and decrease extent of cure. The microwave heating of thermoplastics is challenging until they reach their critical temperature due to their low dielectric loss factor and high degree of crystallinity. Higher the degree of crystallinity, more is the restriction of changes in the dipole orientation in the oscillatory electric field. The energy absorption mechanism by the thermosets and thermoplastics is schematically explained in Figure 4.8. The absorption of microwave energy is dominated by the fibres for the polymer composites with high degree of crystallinity and temperature below the critical temperature. The crystallinity of polymer matrix is more important as amorphous polymers heat more effectively than the semi-crystalline or crystalline materials. The component with high dielectric loss absorbs the microwave radiation, gets heated and transfers the heat to the other component through conduction. For PMCs with low degree of crystallinity, the microwave absorption is dependent on matrix as well as the reinforcement. As the thermoplastic matrix absorbs microwave radiation, the temperature of the matrix reaches beyond its critical temperature. Hence, more uniform curing of PMCs is achieved, in particular with fibre reinforcement of high conductivity. For curing the thermoset-based PMCs, the energy absorption is dominated by energy absorption by the component with high dielectric loss and thermal conductivity of the fibres at elevated temperatures. Further, the anisotropic properties and heterogeneous microstructure of the PMCs significantly affects the microwave and energy absorption. It has also been reported that in case the fibres absorb the microwave energy, increased temperature of fibres results in change in bonding between the fibre and matrix, which results in increased adhesion between the matrix and fibres. Thermoplastics are known to be fully polymerised that melt and flow on application of heat. They are processed above their glass-transition temperature or melting points to reduce the melt viscosity and allow flow to promote adhesion. High-performance, semi-crystalline thermoplastics and polymers, such as polyetheretherketone (PEEK), can be difficult to be processed using microwaves until a critical temperature is reached, beyond which the heating rate increases significantly. The critical temperature is related to increased molecular mobility that may not be the same as the glass-transition temperature of the polymer. Although the PMCs that are candidates for microwave processing are not typically conductive. However, by the presence of particles or fibres that are conductive or having dielectric properties significantly different than the matrix significantly influence the way in which the composite materials interacts with the microwave radiation and may also aid in processing and modifying the mechanical, physical or optical properties (National Research Council 1994). The conducting reinforcement also modifies the electric field pattern compared to the neat resin. Some examples of conductive additives include carbon black, CF, metal fibres and metal flakes. The conductive surfaces of the reinforcements interact strongly with the microwave radiation. The effect of conductive additives on microwave heating and skin depth of composite depends on the shape, size, concentration and electrical resistivity of the inclusions and their distribution in the matrix. It is also reported by many researchers that the presence of conductive fillers may inhibit microwave heating by decreasing the skin depth. However, by controlling the nature orientation, and concentration of the fillers, the microwave response of the material can be tailored over a broad range. For example, CF have a very high resistivity and they can heat the surrounding matrix very efficiently. This preferential heating has shown improved interfacial adhesion between the fibres and matrix, which subsequently leads to improved properties of the microwave-processed PMCs (Agrawal and Drzal 1989). Preferential heating of conductive fillers has also been utilised in joining of polymers and polymeric composites. It has been have reported an increase rate of cross-linking in a composite system due to high dielectric loss due to presence of conductive filler (aluminium powder). Similar results of carbon black-filled epoxy resin system were also reported (National Research Council 1994). Non-conductive additives such as GF and non-conducting metal oxides, which are used as reinforcements (such as oxides), can also influence composite properties through preferential heating mechanisms, depending on their dielectric properties. Microwave processing of PMC's in spite of several processing challenges is time saving and energy efficient and environmentally clean than the conventional processing routes. Not all polymer materials are suitable for microwave processing. However, many polymers contain groups that form strong dipoles (e.g. epoxy, hydroxyl, amino, cyanate). Microwave processing can be used to process a wide range of polymer and products, including thermoplastic and thermosetting resins.

4.7 CHALLENGES AND OPPORTUNITIES IN MICROWAVE PROCESSING OF POLYMER MATRIX COMPOSITES

PMCs are known to have anisotropic properties and heterogeneous distribution of reinforcing fibres in the matrix, which makes them inherently difficult for them to be processed through microwave energy. Non-uniform change in dielectric properties of the matrix and reinforcements with increasing temperature may lead to formation of hot spots that can cause material burn out. The differential microwave absorption rates in different phases may result in poor adhesion between the reinforcement and matrix. The differential heating behaviour of the PMC in microwave processing can be minimised though the use of special tooling and fixture design. Another challenge in microwave processing of PMCs is that processing temperature is often very close to its thermal degradation temperature, making temperature control a critical parameter. If the temperature is too high, the polymer matrix might undergo undesirable cross-linking, scission or oxidation. One or more of these phenomena may cause significant changes in the mechanical properties of these polymer composites. This places strict requirement of uniform temperature distribution throughout the part being processed. Real-time measurement of changing material properties during microwave exposure is difficult. Therefore, modelling and simulation of microwave heating of PMCs have limited scopes, due to unavailability of the valuable experimental data. The hygroscopic nature of PMC's also makes microwave-processing route challenging for these materials. Secondly, the size of the part is limited due to availability of standard size of microwave applicators. The customisation of the applicators demands higher investment. Furthermore, leakage of microwave is harmful for human beings, therefore proper safety precautions must be followed during microwave processing. According to a recent review article by Mishra and Sharma, only 8% of the published research is in the field of polymers and PMC-based materials (Mishra and Sharma 2016). This indicates that still there is lot of scope in researching on processing of polymers and PMCs using microwave energy. The road map of application microwaves in polymer processing can be summarised in Figure 4.9.

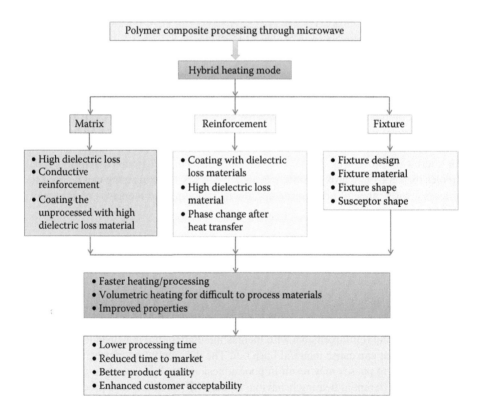

FIGURE 4.9   Road map for application of microwaves in polymer composite processing.

4.8 CONCLUSIONS

The use of microwaves in material processing was identified several years ago; however, due to limited understanding of the phenomena involved in material processing it was largely confined to a few materials. Now, the material processing community has realised the potentials of microwave technology over conventional material-processing routes. With further improvements and optimisation of microwave processing, there are unlimited opportunities with possible innovation in process, fixture and process monitoring. This will entail enhanced understanding in microwave processing of polymer composite materials. There are immense opportunities to use microwaves in material processing, innovative approaches in design and manufacturing of polymer composites materials.

REFERENCES

Agrawal, D. 2010. Latest global developments in microwave materials processing. *Materials Research Innovations* 14 (1): 3–8.

Agrawal, R.J., and L.T. Drzal. 1989. Effects of microwave processing on fiber—Matrix adhesion in composites. *Journal of Adhesion* 29 (1–4): 63–79.

Ahmed, A., and E. Siores. 2001. Microwave joining of 48% alumina–32% zirconia–20% silica ceramics. *Journal of Materials Processing Technology* 118 (1–3): 88–94.

Aravindan, S., and R. Krishnamurthy. 1999. Joining of ceramic composites by microwave heating. *Materials Letters* 38 (4): 245–249.

Budinger, D.E. 2008. Microwave brazing process. 290137, issued 2008.

Chandrasekaran, S., T. Basak, and S. Ramanathan. 2011. Experimental and theoretical investigation on microwave melting of metals. *Journal of Materials Processing Technology* 211 (3): 482–487.

Chawla, K.K. 2012. *Composite Materials: Science and Engineering.* New York: Springer Science & Business Media.

Clark, D.E., D.C. Folz, and J.K. West. 2000. Processing materials with microwave energy. *Materials Science and Engineering: A* 287 (2): 153–158.

Clark, D.E., and W.H. Sutton. 1996. Microwave processing. *Annual Reviews of Materials Science* 26: 299–331.

Fang, C.Y., C.A. Randal, M.T. Lanagan, and D.K. Agrawal. 2008. Microwave processing of electroceramic materials and devices. *Journal of Electroceramics* 22 (1–3): 125–130.

Farag, S., A. Sobhy, C. Akyel, J. Doucet, and J. Chaouki. 2012. Temperature profile prediction within selected materials heated by microwaves at 2.45GHz. *Applied Thermal Engineering* 36 (April): 360–369.

Jones, D.A., T.P. Lelyveld, S.D. Mavrofidis, S.W. Kingman, and N.J. Miles. 2002. Microwave heating applications in environmental engineering—A review. *Resources, Conservation and Recycling* 34 (2): 75–90.

Leonelli, C., P. Veronesi, L. Denti, A. Gatto, and L. Iuliano. 2008. Microwave assisted sintering of green metal parts. *Journal of Materials Processing Technology* 205 (1): 489–496.

Ku, H.S., E. Siores, A. Taube, and J.A.R. Ball. 2002. Productivity improvement through the use of industrial microwave technologies. *Computers & Industrial Engineering* 42 (2–4): 281–290.

Mantia, F.P., and M. Morreale. 2011. Green composites: A brief review. *Composites Part A: Applied Science and Manufacturing* 46 (6): 579–588.

Metaxas, A.C. 1991. Microwave heating. *Power Engineering Journal.* IET Digital Library. doi:10.1049/pe:19910047.

Mishra, R.R., and A.K. Sharma. 2016. Microwave–material interaction phenomena: Heating mechanisms, challenges and opportunities in material processing. *Composites Part A: Applied Science and Manufacturing* 81: 78–97.

Mondal, A., A. Shukla, A. Upadhyaya, and D. Agrawal. 2010. Effect of porosity and particle size on microwave heating of copper. *Science of Sintering* 42 (2): 169–182.

National Research Council. 1994. *Microwave Processing of Materials*. Washington, DC: The National Academies Press.

Oghbaei, M., and O. Mirzaee. 2010. Microwave versus conventional sintering: A review of fundamentals, advantages and applications. *Journal of Alloys and Compounds* 494 (1–2): 175–189.

Ramkumar, J., S. Aravindan, S.K. Malhotra, and R. Krishnamurthy. 2002. Enhancing the metallurgical properties of WC insert (K-20)/cutting tool through microwave treatment. *Materials Letters* 53: 200–204.

Roger, J.M. 1998. *Engineer's Handbook of Industrial Microwave Heating*. The Institution of Engineering and Technology.

Roy, R., D. Agrawal, J. Cheng, and S. Gedevanishvili. 1999. Full sintering of powdered-metal bodies in a microwave field. *Nature* 399 (6737): 668–670.

Sgriccia, N., and M.C. Hawley. 2007. Thermal, morphological, and electrical characterization of microwave processed natural fiber composites. *Composites Science and Technology* 67 (9): 1986–1991.

Sutton, W.H. 1989. Microwave processing of ceramic materials. *American Ceramic Society Bulletin* 68: 376–386.

Thostenson, E.T., and T.-W. Chou. 1999. Microwave processing: Fundamentals and applications. *Composites Part A: Applied Science and Manufacturing* 30 (9): 1055–1071.

Zhu, S., W.G. Fahrenholtz, G.E. Hilmas, S.C. Zhang, E.J. Yadlowsky, and M.D. Keitz. 2008. Microwave sintering of a ZrB2–B4C particulate ceramic composite. *Composites Part A: Applied Science and Manufacturing* 39 (3): 449–453.

5 Fabrication and Stamp Forming of Thermoplastic Composites

S. Suresh and V.S. Senthil Kumar

CONTENTS

5.1 INTRODUCTION

At present, the fibre-reinforced composites with higher specific strength and stiffness, shorter processing time and lower manufacturing cost are the promising materials for lightweight automotive components [1,2]. The thermoplastics are having number of advantages over thermosets and act as an alternate matrix material. However, the major crucial factors such as improper processing techniques, void formation and poor wettability were restricting the applications of these materials. Hence, extensive researches were carried out in improving the fibre/matrix adhesion by various fibre-surface treatments and matrix modifications for increasing the performance of thermoplastic composite materials.

The tensile strength of glass wool-reinforced polypropylene (PP) composites was evaluated and was identified that the silane-treated glass wool enhanced the tensile strength of composites [3]. The effect of a functionalised PP for the enhancement of adhesion strength between matrix PP and silane-treated glass fibre (GF) was studied [4].

The impact of PP chains on the adhesion as well as on the mechanical properties of GF/PP composites was studied [5]. The mechanical performance optimisation over interfacial strength by maleated PP in GF/PP composites was made [6]. The mechanical properties of long GF/PP composites by varying the concentration of dicumyl peroxide and maleic anhydride in compatibilizer were investigated [7]. The effects of matrix modification on the mechanical, morphological and thermal properties of carbon fibre-reinforced PP composites [8–9] as well as different types of natural fibre-reinforced PP composites [10–12] were also studied. Most of the previous studies discussed about the properties of short or long GF-reinforced PP composites. But, very few studies were reported about the continuous fibre and woven fabric composite materials. The present study investigated the influence of coupler concentration on the tensile properties of novel plain-woven glass fibre-reinforced thermoplastic (GFRTP) composites. The morphological study was conducted on the samples using scanning electron microscope (SEM) to analyse the interaction between matrix and fibre. Finally, the optimum coupler concentration was obtained to fabricate the thermoplastic composites with superior mechanical characteristics.

5.2   RAW MATERIALS

An isotactic PP (0.5 mm film) was selected as the matrix material. Novel plain-weave E-glass fibre (280 gsm) fabric structure was selected as the reinforcement material. This fabric structure enhances the degree of impregnation of matrix through the reinforcement than the conventional plain-weave structure. The maleic anhydride grafted PP was selected as the matrix compatibilizer. The (3-aminopropyl) trimethoxy silane (MW 179.29, $H_2N-(CH_2)_3-Si-(OCH_3)_3$) was used as the silane-coupling agent. The silane contains hydrolysable methoxy group that react with inorganic materials and organo-functional amine group that react with organic materials. The raw materials used for the fabrication of GFRTP composites were presented in Figure 5.1. The properties of raw materials [13,14] were listed in Table 5.1.

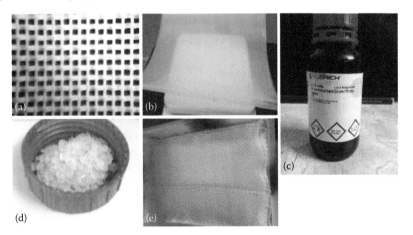

FIGURE 5.1   Raw materials: (a) glass fibre fabric, (b) polypropylene film, (c) (3-aminopropyl) trimethoxy silane, (d) MAPP pellets and (e) fabric in silane solution.

TABLE 5.1
Properties of Raw Materials

Raw Materials	Density (g/cm³)	Tensile Strength (MPa)	Young's Modulus (GPa)
Polypropylene	0.905	33	1.34
Glass fibre	2.55	2400	73

First, the silane-coupling agent with 1.5% concentration was diluted with distilled water. The silane-coupling agent was gradually added while stirring the distilled water solution and the stirring was continued for 20–30 min. The GF fabric was dipped in the hydrolysate solution for 2–3 h. The silanol solution formed during hydrolysis, reacted with GF surfaces and the chemical reaction, as shown in Equation 5.1, took place between them and resulting in hydrogen bonding [5,15]. After that, the fabric was pulled out and dried in the sunlight for a day to remove the excessive liquids and subsequently silane developed a strong covalent bond with the GF surface.

$$\text{GF surface] – Si - OH + (OH)}_3 \text{ – Si - (CH}_2)_3\text{NH}_2 = \text{GF surface] - Si - O - Si - (CH}_2)_3\text{NH}_2 + \text{H}_2\text{O}$$

with $(OH)_2$ above the second Si.

$$(5.1)$$

5.3   HOT-COMPRESSION MOULDING

Numerous techniques were implemented in the manufacturing of thermoplastic composites such as film stacking, thermoforming and vacuum bagging. In the present study, the GFRTP composite laminates were fabricated using film-stacking technique in hot compression-moulding machine. The aminosilane-treated GF fabric was stacked in between PP film for 2 mm thickness and the maleic anhydride grafted polypropylene (MAPP) powder was distributed uniformly in all layers. The stacked materials (300 × 300 mm size) were placed in the hydraulic press and then heated above the melting temperature of the matrix. The forming pressure and forming temperature used were 90 bar and 190°C, respectively. The applied pressure was maintained until the assembly was cooled below the melting temperature in order to avoid the occurrence of voids or defects. In order to study the influence of coupler, the concentration of coupler was varied from 0wt.% to 16wt.% instep of 4wt.%. Table 5.2 shows the composition of constituent materials of the fabricated GFRTP composites.

At the forming temperature, the molten MAPP will react with both silane-treated GF fabric and matrix material. This MAPP co-polymer was physically absorbed into the PP matrix and chemically attached to the organo-functional amine group of silane-treated GF surface. The chemical reaction took place during the high-temperature compression and subsequently the strong interfacial bonding had occurred between the constituent materials. Finally, the assembly was allowed to cool below the melting temperature of the matrix, whereas the applied pressure was maintained to avoid the occurrence of voids or defects within the laminate. After cooling down, the GFRTP composite laminate should be separated from the die frame.

TABLE 5.2
Composition of Constituent Materials

S. No	Sample ID	Glass Fibre Fabric (wt.%)	Polypropylene (wt.%)	Maleic Anhydride-g-Polypropylene (wt.%)
1	GFRTP-0	40	60	0
2	GFRTP-4	40	56	4
3	GFRTP-8	40	52	8
4	GFRTP-12	40	48	12
5	GFRTP-16	40	44	16

5.4 TENSILE TESTING

The tensile test was conducted as per ASTM D 638 Type I standard with the specimen size of 165 × 19 × 2.2 mm. The test was performed to understand the tensile behaviour of composites under uniaxial loading condition and was carried out at room temperature with a crosshead velocity of 2 mm/min using 5-ton universal testing machine (UTM). During tests, an electronic load cell was used to measure the load and a linear variable differential transducer was used to measure the displacement.

The tensile properties of GFRTP composites were presented in Table 5.3. It was perceived that the addition of MAPP in GF/PP interface improved the tensile strength of GFRTP composites from 116 to 139 MPa [3]. Further, the increased load-carrying capacity of composites confirmed the presence of strong bonding between GF and PP. In order to demonstrate the influence of MAPP on fibre/matrix interface, the SEM micrographs of fractured tensile specimens' fibre surfaces were shown in Figure 5.2a–e for different MAPP concentrations. The difference of surface morphology for various GFRTP-composite specimens was due evidently to the nature of interfacial interaction of MAPP compatibilizer over the PP matrix and the silane-treated GF surfaces. The SEM image of GFRTP-0 composite in Figure 5.2a shows the glass-fibre surfaces with small retention of the resin attachment, which confirms

TABLE 5.3
Tensile Properties of GFRTP Composites

Sample ID	Maximum Load (kN)	Elongation (%)	Tensile Strength (MPa)	Tensile Modulus (GPa)
GFRTP-0	3	9.1	116	1.27
GFRTP-4	3.71	9.8	130	1.32
GFRTP-8	3.83	9.3	140	1.50
GFRTP-12	3.14	8.5	105	1.24
GFRTP-16	3	7.9	101	1.28

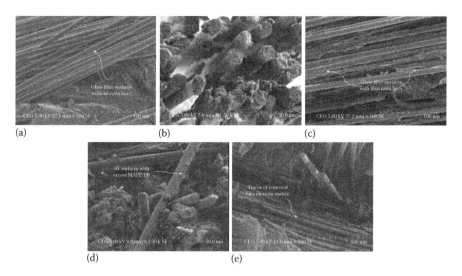

FIGURE 5.2   SEM micrographs of the fracture surfaces of GFRTP composites for different concentrations of MAPP: (a) 0%, (b) 4%, (c) 8%, (d) 12% and (e) 16%.

the weak interfacial adhesion between GF and PP matrix. This subsequently leads to the peeling out of resin matrix from the glass-fibre surfaces. The tensile strength of GFRTP-0 composite was 116 MPa and was increased by 20.69% due to the addition of 8wt.% MAPP. This revealed that the GFRTP composite reached maximum tensile values when MAPP/PP ratio was 15%. The SEM images of GFRTP-4 and GFRTP-8 composites in Figure 5.2b and c indicate the presence of a thin layer of resin matrix throughout the glass-fibre surfaces, even after the failure, which ensures the relatively strong interfacial adhesion between them. Such an improved interfacial adhesion would lead to a higher mechanical performance, which was in good agreement with the above-mentioned mechanical properties [8]. This enhanced adhesion was due greatly to the effective chemical bonding between hydroxyl functional groups with GF surfaces and also organo-functional amine groups with maleic anhydride in MAPP. The flow of molten PP through novel GF fabric gaps greatly improved the effectiveness of in situ impregnation between them, which also contributed for the improvement of tensile properties of GFRTP composites.

The tensile modulus of GFRTP composites differed from 1.24 to 1.5 GPa and reached the maximum value at 8wt.% MAPP concentration. From the results, it was evident that the MAPP concentration had a much insignificant effect on the tensile modulus of composite materials [3]. Both tensile strength and modulus were decreased considerably after 8wt.% MAPP concentration due to the agglomeration of excessive MAPP compatibilizer over the GF/PP interface [8]. The fracture morphology of GFRTP-12 and GFRTP-16 specimens was illustrated in Figure 5.2d and e. This agglomeration was clearly visible in SEM images of GFRTP composites as illustrated in Figure 5.2d. This excessive MAPP compatibilizer starts severely degrading the load transferring capability of PP matrix and subsequently the properties of composites. The traces of removed glass fibres on the PP matrix were indicated in Figure 5.2e.

5.5   STAMP FORMING

Stamp forming is a technique that is widely used for the hemispherical forming of thermoplastic composites. The forming of complex double curvature-shaped components involves various material flow behaviours as well as deformation mechanisms. Hence, lot of studies have been conducted to investigate the forming and deformation behaviour of thermoplastic composites during stamp forming. Sachihiro et al. [16] decided that the draw depth strongly depends on the test temperature and holding pressure during an isothermal forming system. Zhang et al. [17] proposed a new thermal stamping method to fabricate a composite part. The results indicated that the drawing force drops rapidly with the increase of forming temperature in the drawing process. Vanclooster et al. [18] investigated the effect of different fibre orientations on the formability of composite laminates using hemispherical punch. A drastic decrease in formability was noticed when the relative fibre orientation increases. Lee et al. [19] investigated the effect of blank holder force on formed shapes of non-crimp fabrics during stamp forming. Davey et al. [20] developed a finite element model to simulate the stamp forming of carbon fibre-composite sheets and observed that the model accurately simulates the evolution of strain and the deformation behaviour. The key emphasis of the study was to investigate the influence of essential stamping process parameters on the forming behaviour of glass fibre-reinforced thermoplastic composite. The GFRTP-8 composite with higher tensile properties was considered for the hemispherical stamp-forming operations.

5.6   EXPERIMENTAL SETUP AND DESIGN

The stamp-forming experimental setup [21] as shown in Figure 5.3 consists of an oven used for heating the blank, the stamping die-set, a support frame used to transfer the blank to the die, the temperature control unit used to heat the die to the

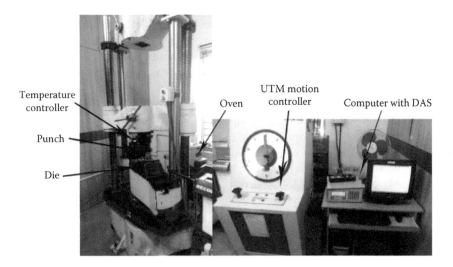

FIGURE 5.3   Stamp-forming experiment setup.

TABLE 5.4
Process Parameters and Their Levels

			Level 1	Level 2	Level 3
S. No.	Process Parameters	Unit	−1	0	+1
1	Die temperature	°C	30	100	170
2	Blank temperature	°C	170	200	230
3	Blank holder force	kN	2	5	8

required temperature and the computer with data acquisition system. The stamping die-set consists of a hemispherical punch, a spring-loaded blank holder and a die. The 27.5 mm radius punch and 30 mm radius die with heating unit were mounted on the 100-ton UTM. The hemispherical punch was used to take the advantage of applying a uniform load in all directions of the blank.

The stamp-forming experiments were designed using response surface methodology's (RSM's) Box–Behnken design, which helps in reducing the number of experimental runs. The three key stamping-process parameters selected for the investigation were die temperature, blank temperature and blank holder force. The levels of parameters were selected based on the literature survey [20,22,23]. The stamping-process parameters and their levels were listed in Table 5.4.

Before the stamp-forming operation, each circular blank was pre-heated using the oven to the specified temperature and then transferred to the die-using frame within 5 s to avoid the heat loss of the blank. Teflon films were placed below the circular blank to reduce the friction and to avoid the sticking of molten matrix to the contact surfaces. The stamping experiments were carried out according to RSM's Box–Behnken design for the various combinations of process parameters using the stamp forming setup. During stamping process, the designed blank holder force was applied on the blank to impose a tension onto it, which helped to reduce the occurrence of wrinkles. Then, a suitable draw force was applied by the hemispherical punch to deform the blank to the desired double-curvature dome shape.

In this study, the forming ratio and true thickness strain were the quality characteristics used to assess the influence of parameters on the stamp-formed components. The draw depth and part thickness at flange were measured by using the digital vernier depth gauge and the coordinate-measuring machine. To assess the forming behaviour effectively, the responses were converted into useful quality characteristics using the relations (2) and (3). Draw force during the stamp-forming process was recorded using the load cell attached in the UTM.

The forming ratio (A_1/A_2), which was the ratio of area of blank to the area of formed dome shape, plays an important role in a composite sheet forming [24].

$$\text{Forming ratio} = \frac{\text{Area of blank}}{\text{Area of formed dome shape}} = \frac{\pi r_0^2}{2\pi r_{\text{dome}} h} = \frac{r_0^2}{2 r_{\text{dome}} h} \quad (5.2)$$

The standard forming ratio for the full-depth stamp formed component was obtained as 1.68 by taking the height of the dome as equal to the die radius.

TABLE 5.5

Responses and Quality Characteristics

S. No	Die Temperature (°C)	Blank Temperature (°C)	Blank Holder Force (kN)	Draw Depth (mm)	Part Thickness (mm)	Forming Ratio	True Thickness Strain	Draw Force (kN)
1	30	170	5	26.00	2.49	1.94	0.22	7.98
2	170	170	5	28.24	1.94	1.79	−0.03	1.64
3	30	230	5	26.16	2.55	1.93	0.24	7.66
4	170	230	5	28.16	1.91	1.79	−0.05	1.72
5	30	200	2	27.00	2.67	1.87	0.29	6.91
6	170	200	2	29.25	1.96	1.72	−0.02	1.44
7	30	200	8	25.36	2.48	1.99	0.22	9.62
8	170	200	8	29.15	1.78	1.73	−0.12	2.32
9	100	170	2	27.72	2.73	1.82	0.31	5.56
10	100	230	2	27.95	2.69	1.80	0.30	4.05
11	100	170	8	26.54	2.46	1.90	0.21	6.22
12	100	230	8	27.50	2.30	1.83	0.14	5.39
13	100	200	5	27.60	2.42	1.83	0.19	5.00
14	100	200	5	27.65	2.43	1.82	0.19	5.02
15	100	200	5	27.55	2.39	1.83	0.18	4.90

The part thickness was measured at the flange approximately at a distance of 5 mm along warp and weft directions from the hemispherical surface to avoid the measurement error. From the part thickness, the true thickness strain was calculated to evaluate the amount of wrinkling and thinning occurred on the flange. The true thickness strain was calculated by using the relation (Equation 5.3) [22]. This was done to clearly show the difference between the actual part thicknesses compared to that of the original blank thickness.

$$\text{True thickness strain } (\varepsilon_t) = \ln\left(\frac{t_{\text{avg}}}{t_0}\right) \quad\quad\quad (5.3)$$

The calculated values of forming ratio as well as the true thickness strain were also tabulated in Table 5.5. The negative sign of true thickness strain indicates the thinning and folding, whereas the positive sign indicates the wrinkling and de-consolidation of the stamp formed components at the flange of the fibre orientations.

5.7   RESPONSE-SURFACE METHODOLOGY

RSM is a set of mathematical and statistical techniques helpful for the analysis and modelling of problems in which responses were influenced by many parameters and the aim was to develop the mathematical models for the quality characteristics, which would be expressed as the function of die temperature, blank temperature and blank holder force. Final Equations 5.4 and 5.5 in terms of actual factors were as follows:

$$\text{Forming ratio} = +1.987 - 1.237e - 03 * DT - 3.669e - 04 * BT$$
$$+ 9.877e - 03 * BHF \qquad (5.4)$$

$$\text{True thickness strain} = + 0.328 + 2.903e - 03 * DT - 3.074e - 04 * BT$$
$$- 0.018 * BHF - 2.505e - 05 * DT^2 \qquad (5.5)$$

5.7.1 VALIDATION OF THE DEVELOPED MODEL

Finally, for verifying the prediction ability of the response surface model, the confirmation tests were conducted randomly for few test conditions as listed in Table 5.6. The experimental quality characteristic values and their corresponding predicted values obtained through Equations 5.4 and 5.5 are presented in Table 5.6. The confirmation tests show that the predicted values were very close to the experimental results and the percentage error lie within the permissible limits.

5.7.2 ANOVA RESULTS

The purpose of ANOVA is to investigate the most significant factors that affect the quality characteristics. This analysis was carried out for the confidence level of 95% and the value of 'α' selected is 0.05. F-ratio was an index used to check the adequacy of the model. From the Tables 5.7 and 5.8, it was apparent that, the calculated F-values were all greater than the F-table value ($F_{0.05,2,8}$ = 4.46) and hence the developed response model was quite adequate. In this investigation, die temperature was found to be the most dominant factor playing a major role followed by a blank holding force; whereas the effect of blank temperature was found to be insignificant.

TABLE 5.6
Confirmation Test Results

Experiment No.	Die Temperature (°C)	Blank Temperature (°C)	Blank Holder Force (kN)	Quality Characteristics (Actual)		Quality Characteristics (Predicted)	
				Forming Ratio	True Thickness Strain	Forming Ratio	True Thickness Strain
1	30	200	5	1.93	0.26	1.93	0.24
2	30	230	8	1.96	0.22	1.94	0.18
3	100	200	2	1.79	0.28	1.81	0.27
4	170	170	8	1.74	−0.12	1.79	−0.10
5	170	200	5	1.76	−0.06	1.75	−0.05

TABLE 5.7
ANOVA for Forming Ratio

Factor	Degrees of Freedom	Sum of Squares	Mean Squares	F-ratio	Contribution %
DT	2	0.0618	0.0309	39.29	77.94
BT	2	0.0032	0.0016	2.04	4.04
BHF	2	0.0080	0.0040	5.09	10.09
Error	8	0.0063	0.0008		7.93
Total	14	0.0793			100.00

TABLE 5.8
ANOVA for True Thickness Strain

Factor	Degrees of Freedom	Sum of Squares	Mean Squares	F-ratio	Contribution %
DT	2	0.2339	0.1169	619.28	87.40
BT	2	0.0053	0.0026	13.98	1.97
BHF	2	0.0269	0.0135	71.31	10.06
Error	8	0.0015	0.0002		0.56
Total	14	0.2676			100.00

5.8   INFLUENCE OF PROCESS PARAMETERS

5.8.1   INFLUENCE OF PARAMETERS ON FORMING RATIO

The influence of process parameters on the forming ratio can be seen in Figure 5.4. The experiments 2, 4, 6 and 8 show the forming ratio obtained at high die temperature condition. These forming ratio values were closer to the standard forming ratio of 1.68, which showed the higher draw depth obtained by the composite blanks at high die temperature condition. This was because of the reduced friction between the heated blank and the hot die. During stamp forming, the blank heated at 170°C approaches

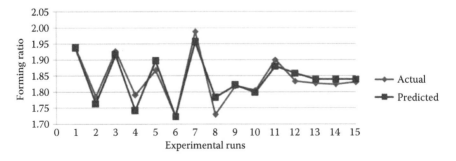

FIGURE 5.4   Effect of process parameters on forming ratio.

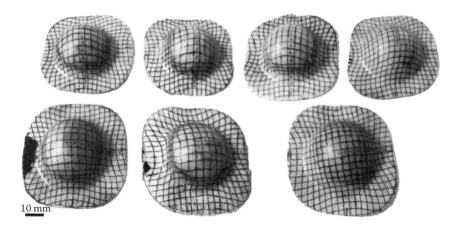

FIGURE 5.5 Few stamp-formed hemispherical components.

the crystallisation temperature at a faster rate; hence, the viscosity of the blank gets increased which leads to the reduction in draw depth. On the other hand, when the blank is at 230°C, the fluidity of the resin matrix enhances, which allows the free movement of the reinforcement, but the matrix is subjected to oxidation, which in turn decreases the quality of the profile obtained.

The experiments 1, 3, 5 and 7 show the forming ratio obtained at low die temperature condition. In these cases, the forming ratio obtained approaches 1.95, which is far away from the standard value. The experimental observations had shown that the composite blanks formed at low die temperature condition experience high friction due to high temperature gradient, which in turn reduces the draw depth of GFRTP composites. This phenomenon was also confirmed by the high draw force needed for pulling the composite blanks into the die cavity against the friction at low die temperature condition as shown in Table 5.5. Figure 5.5 shows few stamp-formed hemispherical components [21]. The rest of the experiments in Figure 5.4 were carried out at medium die temperature condition. In these cases, the forming ratios lie between the above two cases. Hence, the die temperature acts as the most predominant stamp-forming parameter in controlling the forming behaviour of thermoplastic composites [25].

Blank holder force creates a desirable friction between heated blank and the blank holder, which opposes the flow of blank and introduces shearing and stretching forces [20]. Observation of composite blanks formed at high and low blank holder force conditions as in Table 5.5 provides an indication of the influence of blank holder force on the forming ratio of the thermoplastic composite materials. As expected, the composite blank under low blank holder force formed at low die temperature condition experiences severe wrinkling on the flange region of fibre directions. However, the same blank but formed at high die temperature experiences out-of-plane buckling on the die shoulder as well as thinning and folding on the flange regions of fibre directions, respectively. From the experimental observations, it was noticed that the increase in blank holder force gradually improves the quality of the profile obtained by the control flow of composite

blank into the die cavity and prevents buckling phenomenon. The predicted forming ratios, for all the designed experiments, obtained by the mathematical model
have very good agreement with the actual experimental forming ratios and are
illustrated in Figure 5.4.

5.8.2 INFLUENCE OF PARAMETERS ON TRUE THICKNESS STRAIN

The true thickness strain was calculated to clearly show the difference between
the thicknesses obtained compared to the actual thickness of the composite
blank. The influence of process parameters on true thickness strain can be seen in
Figure 5.6. As expected, the wrinkling obtained on the flange region in tests 1, 3,
5 and 7 produce a higher thickness strain compared to that of tests 2, 4, 6 and 8.
As mentioned earlier, when composite blank was formed at a low die temperature
(30°C), the increase in resin viscosity combined with the re-orientation of fibres
will generate an over part thickness at the flange. This phenomenon of thicken-
ing at the flange also occurs due to the insufficient time for the heated blank to
redistribute uniformly before it reaches its crystallisation temperature. Hence, the
partially consolidated blanks produce wrinkling at the flange area. On the other
hand, the composite blank formed at high die temperature (170°C) shows the
thinning on the flange of the stamp-formed components due to the easy flow of
high fluidity resin matrix.

Experimental observations of the tests 5, 9 and 10 exhibit higher true thickness
strain values due to the blank, which was formed at low blank holder force. But
the composite blanks formed at high blank holder force produce the stamp-formed
components without any wrinkling due to the controlled movement of the reinforcement into the die cavity. The effect of blank temperature on thickness strain
was very negligible due to the strong influence of the other two parameters. As
shown in Figure 5.6, the predicted true thickness strains obtained by the mathematical model had a very good agreement with the actual experimental true thickness strain values.

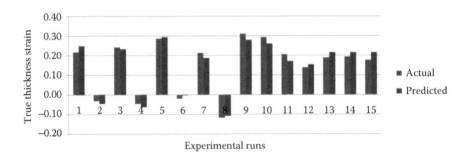

FIGURE 5.6   Effect of process parameters on true thickness strain.

5.9   CONCLUSION

The glass fibre-reinforced PP composites were fabricated by using the film-stacking method in the hot compression-moulding machine. The influence of MAPP concentration on morphological and tensile properties of aminosilane-treated glass fibre-reinforced PP composites was investigated. As well as the influence of process parameters on the hemispherical forming were also investigated. The conclusions from the present study were as follows:

- In general, the mechanical properties of GFRTP composites were improved considerably by the addition of MAPP over GF/PP interface. However, the drop in mechanical properties was observed after the optimum MAPP concentration.
- The GFRTP composite was obtained with better performance characteristics when the ratio of MAPP/PP was 15%. The MAPP concentration increased the conventional tensile strength and tensile modulus by 20.69% and 18% respectively.
- The SEM images confirmed that the improvement in mechanical properties was due evidently to the improved interfacial adhesion between GF fabric and PP matrix.
- From the test results, it was revealed that the developed GFRTP composite laminate could well be suited for the automotive applications such as door module carrier, dashboard carrier and structural carriers.
- The die temperature acts as a prominent parameter, followed by blank holder force and blank temperature in controlling the quality characteristics of the stamp-formed GFRTP components. It is revealed that the increase in die temperature improves the forming ratio at a larger extent by reducing the friction between the heated blank and the die, but decreases the true thickness strain as well as the draw force.
- It recognised that the increase in blank holder force increases the draw force significantly, but decreases the thickness strain as well as the forming ratio and simultaneously improves the quality of the profile obtained. It is known that the increase in blank temperature improves the forming ratio slightly, but decreases the thickness strain as well as the draw force.
- From this study, it was concluded that the stamp-formed GFRTP components can be obtained with better quality characteristics, when the process parameters lie at the die temperature of 170°C, blank temperature of 200°C and blank holder force of 8 kN condition.
- Stamp-forming operation gives the stamp-formed GFRTP component with the maximum draw depth of 29.15 mm, part thickness of 1.78 mm and draw force of 2.315 kN without any wrinkling and defects.
- The developed regression mathematical models for the quality characteristics were having very good agreement with the experimental output values.

NOMENCLATURE

r_0 – radius of the initial composite blank
r_{dome} – base radius of the dome surface
h – height of the dome surface
t_{avg} – average part thickness
t_0 – the original blank thickness
DT – die temperature
BT – blank temperature
BHF – blank holder force
ε_t – true thickness strain
A_1 – area of blank
A_2 – area of formed dome shape

ACKNOWLEDGEMENT

The authors thank Anna University-Chennai, Velammal Engineering College, METMECH Engineers, Chennai, India for manufacturing and testing of samples.

REFERENCES

1. Yanagimoto J, Ikeuchi K, Sheet forming process of carbon fiber reinforced plastics for lightweight parts, *CIRP Annals - Manufacturing Technology*, 61(1), 2012, 247–250.
2. Hufenbach W, Böhma R, Thieme M et al., Polypropylene/glass fibre 3D-textile reinforced composites for automotive applications, *Materials and Design*, 32, 2011, 1468–1476.
3. Tsukamoto M, Murakami T, Yoshimura Y et al., Evaluation of the tensile strength of polypropylene-based composites containing glass wool, *Materials Letters*, 132, 2014, 267–269.
4. Nygard P, Redford K, Gustafson CG. Interfacial strength in glass fibre-polypropylene composites: Influence of chemical bonding and physical entanglement, *Composite Interfaces*, 9(4), 2002, 365–388.
5. Etcheverry M, Barbosa SE. Glass fiber reinforced polypropylene mechanical properties enhancement by adhesion improvement, *Materials*, 5, 2012, 1084–1113.
6. Sorrentino L, Simeoli G, Iannace S et al., Mechanical performance optimization through interface strength gradation in PP/glass fibre reinforced composites, *Composites Part B*, 76, 2015, 201–208.
7. Zhang K, Guo Q, Zhang D et al., Mechanical properties and morphology of polypropylene/polypropylene-g-maleic anhydride/long glass fiber composites, *Journal of Macromolecular Science Part B*, 54(3), 2015, 286–294.
8. Liu Y, Zhang X, Song C et al., An effective surface modification of carbon fiber for improving the interfacial adhesion of polypropylene composites, *Materials and Design*, 88, 2015, 810–819.
9. Karsli NG, Aytac A. Effects of maleated polypropylene on the morphology, thermal and mechanical properties of short carbon fiber reinforced polypropylene composites, *Materials and Design*, 32, 2011, 4069–4073.
10. Mahmud Zuhudi NZ, Jayaraman K, Lin RJT, Mechanical, thermal and instrumented impact properties of bamboo fabric-reinforced polypropylene composites, *Polymers and Polymer Composites*, 24(9), 2016, 755–766.

11. Aridi NAM, Sapuan SM, Zainudin ES. Mechanical and morphological properties of injection-molded rice husk polypropylene composites, *International Journal of Polymer Analysis and Characterization*, 21(4), 2016, 305–313.
12. Ravi Kumar N, Ranga Rao CH, Srikant P et al., Mechanical properties of corn fiber reinforced polypropylene composites using Taguchi method, *Materials Today: Proceedings*, 2, 2015, 3084–3092.
13. Elmarakbi A, Ed., *Advanced Composite Materials for Automotive Applications Structural Integrity and Crashworthiness*, University of Sunderland, Chichester, United Kingdom: John Wiley & Sons, p. 18, 2014.
14. Boedeker Plastics: Polypropylene Datasheet. Boedeker Plastics, Inc., 904 West 6th Street, Shiner, Texas 77984 USA. http://www.boedeker.com/polyp_p.htm.
15. Arkles B, *Silane Coupling Agents: Connecting Across Boundaries*. 3rd ed. Morrisville, PA: Gelest, 2014.
16. Isogawa S, Aoki H, Tejima M, Isothermal forming of CFRTP sheet by penetration of hemispherical punch, *Procedia Engineering*, 81, 2014, 1620–1626.
17. Zhang Q, Cai J, Gao Q, Simulation and experimental study on thermal deep drawing of carbon fiber woven composites, *Journal of Materials Processing Technology*, 214, 2014, 802–810.
18. Vanclooster K, Lomov SV, Verpoest I, On the formability of multi-layered fabric composites, *Proceedings of 17th International Conference on Composite Materials (ICCM17), 27th–31st Jul 2009, London: Organised by The Institute of Materials, Minerals and Mining.* 2010.
19. Lee JS, Hong SJ, Yu W-R, Kang TJ, The effect of blank holder force on the stamp forming behavior of non-crimp fabric with a chain stitch, *Composites Science and Technology*, 67, 2007, 357–366.
20. Davey S, Das R, Cantwell WJ, Kalyanasundaram S, Forming studies of carbon fibre composite sheets in dome forming processes, *Composite Structures*, 97, 2013, 310–316.
21. Suresh S, Senthil Kumar VS, Influence of process parameters on the forming behavior of thermoplastic composite: An experimental approach, *International Journal of Printing, Packaging & Allied Sciences*, 4(3), 2016, 1553–1564.
22. Sadighi M, Rabizadeh E, Kermansaravi F, Effects of laminate sequencing on thermoforming of thermoplastic matrix composites, *Journal of Materials Processing Technology*, 201, 2008, 725–730.
23. Venkatesan S, Kalyanasundaram S, Finite element analysis and optimization of process parameters during stamp forming of composite materials, *IOP Conference Series: Materials Science and Engineering*, 10, 2010, 012138.
24. Bhattacharyya D, Bowis M, Jayaraman K, Thermoforming woodfibre–polypropylene composite sheets, *Composites Science and Technology*, 63, 2003, 353–365.
25. Lessard H, Lebrun G, Benkaddour A, Pham X-T, Influence of process parameters on the thermostamping of a $[0/90]_{12}$ carbon/polyether ether ketone laminate, *Composites: Part A*, 70, 2015, 59–68.

6 Non-Destructive Testing Methods of Polymer Matrix Composites

Vijay Chaudhary, Pramendra Kumar Bajpai and Sachin Maheshwari

CONTENTS

6.1   INTRODUCTION

The applications of polymer-based composite materials are growing rapidly due to their numerous properties such as low weight, low density, high specific strength, high stiffness and good thermal stability. These substantial properties increase the exhaustive consumption of polymer-based composites materials in the arena of various manufacturing sectors and other industries. Industries such as automobile, aerospace, shipping, textile, construction and building have potential applications of these composites for the fabrication of different parts. In automobile industries,

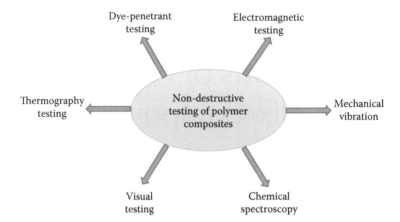

FIGURE 6.1   Non-destructive testing methods for polymer composites.

various parts such as door panels, head liners, dash boards and so on are being fabricated by polymer composites and these parts are assembled together to develop a final product (Rathnakar and Pandian 2015). In aerospace industries, structural mass of an aircraft have potential consumption of polymer composite materials, especially carbon-reinforced polymer composites (Soutis 2015). In building and construction, various parts such as roofs, doors, windows and so on are being manufactured by polymer-based composite materials (Milwich 2010). Besides all the positive properties, there may be some defects present in the polymer-based composite materials. Assembling of various small parts to fabricate final products causes the defects such as cracking in the final product. Some defects can also arise in the processing phase of polymer composites in the form of voids, gas entrapment and porosity. Defects arose at the processing phase and at the assembling time weaken the developed composite part. These defects in the polymer-composite material introduce the various testing methods, which may be used to detect the defects in the developed composite material. Figure 6.1 shows the schematic diagram of various non-destructive techniques used to detect the defects in polymer-based composites.

The testing methods can be broadly categorised in the destructive (Pull-out, microbond, micro-compression and fragmentation) and non-destructive methods (vibration damping, dye penetrant, acoustic emission, thermography, ultrasonic, spectroscopy etc.). Non-destructive testing (NDT) can further be categorised on the basis of contact (traditional ultrasonic testing, eddy current testing, magnetic testing, dye penetrant etc.) and non-contact methods (visual inspection, thermography, radiography, shearography etc.). In contact methods, the testing machine and sensors make good contact with the specimen but in non-contact methods of non-destructive testing, there is no contact between the sensors and the specimen (Gholizadeh 2016). Various authors carried out their studies on the non-destructive testing of polymer composites. Figure 6.2 shows the detailed classification of various methods used in electromagnetic testing, chemical spectroscopy and mechanical vibration non-destructive testing methods.

Imielinska et al. (2004) investigated the impact response (damage due to impact) in carbon/epoxy and carbon/Kevlar/epoxy hybrid composites using air-coupled

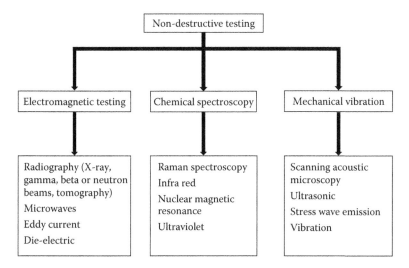

FIGURE 6.2   Methods used in various non-destructive testing methods.

ultrasonic C-scan ultrasonic technique and X-ray radiography technique. The author found that the damage zone and defect size estimation can be found out correctly by both the techniques. Schilling et al. (2005) investigated the internal damage (internal geometry flaws, delamination, matrix cracking and micro-cracking) in glass/epoxy and graphite/epoxy composite using X-ray computed micro-tomography technique. The author concluded that X-ray tomography technique found internal damage of fibre-reinforced polymer composites.

The present study mainly focuses on the various non-destructive testing methods used for the evaluation of damages and defects present in the polymer-based composite materials.

6.2   NON-DESTRUCTIVE TESTING

Non-destructive testing is an inspection process to evaluate the defects present in the interior or at the outer surfaces of the materials without altering the physical and chemical properties of the parent material. NDT provides a very effective method to control the quality of the product by checking the whole product without damaging the material. Various non-destructive techniques such as dye penetrant technique, visual inspection, shearography, magnetic testing, liquid penetrant testing, radiography, acoustic emission, thermal NDT methods, optical methods, vibration damping techniques, chemical spectroscopy are being used to investigate the damage pattern/mechanism of polymer-based composite materials (Gholizadeh 2016). Various authors reported about the non-destructive testing of polymer composites and investigated the different types of defects in the internal geometry of the polymer composites and on the outer surfaces of the composite materials. Table 6.1 shows the different types of defects in the polymer composites investigated by various non-destructive testing methods.

TABLE 6.1

Literature of Various NDT Methods Used for Detection of Defects in Polymer Composites

Types of Defect	Non-Destructive Testing Method	Reference
Voids, delamination and mechanical damage (matrix cracking)	Terahertz radiation	Yakovlev et al. (2014)
Internal defects in composite materials	Automatic ultrasonic radiation	Orazioa et al. (2008)
Cracks and crazes in fibre-based polymer composites	Resonant vibration method	Pye and Adams (1981)
Disbondings in sandwich plates of polymer composites	Holographic non-destructive testing	Jesnitzer and Winkler (1981)
Damage and internal flaws (delamination and micro-cracking)	X-ray computed micro-tomography	Schilling et al. (2005)
Impact damage in thin carbon/epoxy composite	Air-coupled ultrasonic C-scan technique	Lmielinska et al. (2004)
Low-energy impact damage	Eddy current technique	Gros (1995)
Matrix porosity	Ultrasonic wave velocity	Reynolds and Wilkinson (1978)

Pollock (1969) used the stress-wave emission method to monitor the growth of cracks and flaws in the composite materials. The author concluded that stress-wave emission method is also useful for examining the fibre–matrix interfacial adhesion. Cousins and Markham (1977) discussed about the ultrasonic spectroscopy technique used for the characterisation of defects such as depth of delamination in the fibre-reinforced polymers. Cuadra et al. (2013) investigated the damage detection in glass fibre (GF)-reinforced polymer composite using hybrid (combination of acoustic emission, digital image correlation and infrared thermography) non-destructive testing method. The GF-reinforced polymer composite used in their study was subjected to both tensile and fatigue loading conditions. Yashiro et al. (2007) introduced a non-destructive testing technique for composite laminates using visualised lamb-wave propagation. The author concluded that lamb wave propagation technique is very useful for disbonding detection between fibre and matrix phase of composite materials, detection of internal cracks and impact-induced delamination. The various non-destructive testing methods are discussed in the following sections in detail.

6.3   ULTRASONIC TESTING

In the past few decades, ultrasonic testing got well established and is being widely used for non-destructive testing of polymer composite materials for the detection of the internal defects induced during the time of manufacturing phase and at the time of maintenance of product without causing damage to the polymer composite (Renoylds 1985).

There are various ways in which ultrasonic testing of polymer composites can be done. A commonly used ultrasonic testing of polymer composites can be applied using pulse-echo or pulsed through-transmission system with a C-scan transmission

(Stone and Clarke 1975). Various authors performed ultrasonic testing of polymer composites to identify the presence of defects in the developed composites. Bergant et al. (2013) performed the pulse-echo ultrasonic C-scan method to examine the processing defects in the glass/epoxy composites. The authors concluded that C-scan images helps to identify the internal flaws and porosity present in the developed composite layers. El-Sabbagh et al. (2013) investigated the fibre distribution homogeneity in the flax/polypropylene (PP) and glass/PP composite using ultrasonic testing method. Sun and Zhou (2016) investigated the internal delamination process in the layered carbon/epoxy composite using ultrasonic testing method. The authors concluded that the C-scan images of the composite obtained from the ultrasonic test machine helps to detect the internal delamination in the layers of the composite materials.

6.4 THROUGH-TRANSMISSION METHOD

This non-destructive testing method used for the detection of flaws in the polymer composites consists of a transducer, transmitter probe and receiver probe. The test specimen is first immersed into the water to create an acoustic coupling between the transducer and the test specimen. The transmitter probe transmits the ultrasonic wave signals that fall on the test specimen and then moves towards the other side of the test specimen. Receiver probe receives the signal waves, which is coming through the test specimen. Any defects in the polymer composite material in the form of cracks, air bubble, gas entrapment, voids and porosity causes the internal reflection of wave signals and that signal is not received by the receiver probe. Then, the disturbed signals of ultrasonic waves or the attenuated ultrasonic waves transmitted though the composite materials confirm the presence of defects, voids and cracks in the composite materials. Figure 6.3 represents the schematic diagram of non-destructive testing of polymer composite specimen through transmission test method of ultrasonic testing (Ostwal and Sawant 2014).

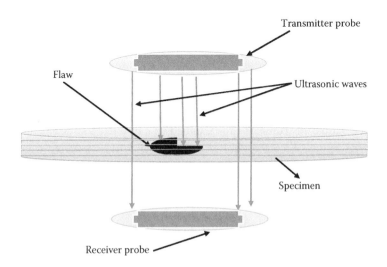

FIGURE 6.3 Through transmission-test method.

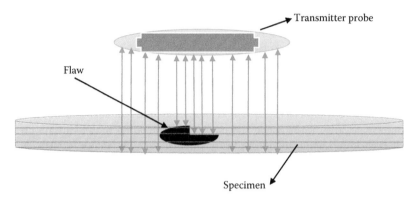

FIGURE 6.4   Pulse-echo test method.

6.5   PULSE-ECHO METHOD

Pulse-echo method of ultrasonic testing is a commonly used non-destructive testing method for the measurement of the internal or external geometrical defects present in the polymer composite materials. In this testing method, a pulse of ultrasonic energy is transmitted through the transmitter probe that falls on the specimen in a direction normal to the surface. After striking the specimen, the ultrasonic waves return from the matrix-reinforced boundaries and from the portion of the composite material which has flaws, voids and cracks. The travel time of ultrasonic waves, transmitted and travelled back from the specimen is recorded. The total travel time is used to identify the portion, position and size of flaws (Stone and Clarke 1975). Figure 6.4 represents the non-destructive testing of polymer composite specimen using pulse-echo test method of ultrasonic testing.

6.6   VISUAL INSPECTION

Visual inspection of polymer composites is the simple and very cheap non-destructive testing method to examine the geometry of polymer composites and detect the presence of flaws in the material. In this test, composite material is investigated without touching the composite specimen and the execution of test is relatively very easy. The accuracy of the test depends on the operator's experience. Many optical instruments are designed to carry out the visual inspection more reliably, often called optical test instruments. Unnporsson et al. materials confirm the presence of defects carried out the visual inspection testing methods of carbon fibre (CF)-reinforced polymer composites. The authors discussed about the digital speckle-photography and electronic speckle pattern interferometry methods of visual inspection technique to identify the surface defects present in the composite material. The following two methods are widely used for the detection of flaws by the visual inspection of polymer composite materials.

6.6.1   DIGITAL SPECKLE PHOTOGRAPHY

Digital speckle photography (DSP) testing method of visual inspection uses the charge-coupled devices (CCD) silicon chips, which is a major technology in digital imaging. P-doped metal oxide semiconductors (MOS) capacitor is used to represent the pixels by the CCD image sensors. During the image acquisition process, these capacitors are biased above the threshold, which convert the incoming photons into the electron charges at the semiconductor oxide interface (Melin et al. 2002). For the image generation in the DSP, two cameras are used at different angles with respect to the composite material specimen. In this way, the displacement in three dimensions can be obtained. First, at the start of monitoring, an image is taken then it is subtracted from all subsequent images. This will make a pattern, used to detect changes in the dimensions (Olsson et al. 2003).

6.6.2   ELECTRONIC SPECKLE PATTERN INTERFEROMETRY

Electronic speckle pattern interferometry (ESPI) testing method of visual inspection also uses the CCD silicon chips discussed in the DSP-testing method. In this testing method, the pattern is generated using a source of laser light. When the light is projected at the composite specimen, the reflection of the laser generates a pattern. To analyse these images, various processing algorithms can be used (Icardi 2003).

These methods are used to detect the damages of the surface in its early stage. Thus, the use of optical methods of visual inspection is limited for the inspection of outer surface of the polymer composites.

6.7   DYE-PENETRATION TESTING

Dye-penetration testing, also known as liquid-penetration testing is the most frequently used non-destructive technique to investigate the surface defects such as cracks, surface porosity, leakage and so on in polymer-based composites. In this technique, phenomenon of capillary action is used by the fluid to penetrate into the surface defects present in the test specimen. First, a liquid fluorescent having low surface tension is applied on the surface by dipping, spraying or brushing. After some time of the application of liquid fluorescent, the excess penetrant is removed from the surface and a developer is deposited. After the application of developer, the invisible flaws present on the test specimen surface become visible. The location of the flaws can be inspected by drawing the penetrant out of the flaw with the help of the developer. These flaws can be inspected by flashing various lights such as ultraviolet, white lights and so on depending on the type of fluorescent used on the surface of test specimen. Figure 6.5 shows the different stages used to identify the presence of defects in the polymer-based composite materials using dye penetrant non-destructive testing method.

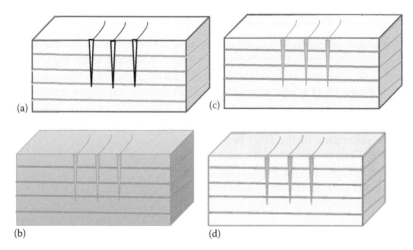

FIGURE 6.5 Dye-penetration technique: (a) test specimen with defects, (b) test specimen dipped in the fluorescent liquid, (c) removal of excess penetrant and (d) application of developer to identify the invisible cracks.

6.8   RADIOGRAPHY TESTING

Radiography testing is a non-destructive testing method to characterise the internal structure and presence of flaws in the polymer-based composite material. In this method, electromagnetic radiation of X-rays and neutrons are used to examine the internal structure and defects in the composite materials. Many authors carried out X-ray radiography testing in their study to evaluate the structure of the developed polymer composite and to identify the presence of defects and flaws in the geometry of the composite material. Prade et al. (2017) characterised the fibre orientation in GF-reinforced polymer and CF-reinforced polymer composites using X-ray radiography. The authors concluded that X-ray radiography-testing method does not restrict the size of the test specimen and allows for very shorter measurement. Schilling et al. (2005) investigated the internal flaws of S-glass/epoxy and graphite/epoxy composite using X-ray micro-tomography. The authors concluded that X-ray micro-tomography can clearly identify the internal flaws such as matrix cracking, fibre fracture and so on in the fibre-reinforced polymer composite. The X-ray radiography-testing method is discussed in the following section in detail.

6.9   CONVENTIONAL X-RAY RADIOGRAPHY METHOD

In this inspection technique, X-rays are generated from the X-ray source, which travels towards the test specimen and strikes it. A radiation-sensitive target, usually a thin film is placed beneath the test specimen. The unabsorbed radiation passing through the voids and defects strike the radiation-sensitive target and thus producing an image that is a two-dimensional shadow image of the test specimen. The various shades of grey colour are developed on the film by absorbing some radiation in

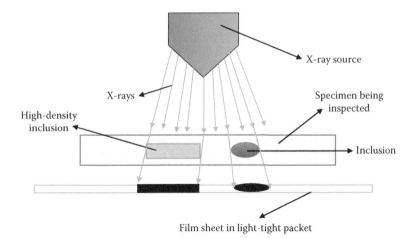

FIGURE 6.6   Conventional X-ray radiography.

the specimen and unabsorbed radiation hits the thin film. This thin film works as a radiogram. This radiography is evaluated by comparing the difference in the density produced by an unknown specimen and the density produced by a similar specimen of high or acceptable quality. Figure 6.6 explains the working principle of conventional X-ray radiography.

6.9.1   X-Ray Computed Tomography

In this technique, a 3D image of component is generated by taking around 360–720 radiograph by rotating the object about one axis. The resulting data are used for the generation of 3D image of the component using the software. The 3D image of the component can be analysed at any plane to detect the flaws.

6.9.2   Enhanced X-Ray Radiography Inspection

The flaws that do not have the sufficient depth in the direction of X-ray beam cannot be detected by conventional X-ray radiography-testing method. To overcome this problem, enhanced X-ray radiography inspection technique is widely used to detect the micro-defects present in the internal structure of polymer composites. The enhanced X-ray radiography uses a contrast medium, generally a radio opaque penetrant. These penetrants show high X-ray absorption coefficient; hence, increase the density of grey shade. In this manner, smaller or micro-flaws can be detected.

6.10   THERMOGRAPHY TESTING

Thermography inspection technique is the most widely used non-destructive testing method to identify the internal defects or flaws by the application of heat on the surface of the test specimen of polymer composite material. In this testing method, the pattern of heat flow is analysed to determine the presence of flaws in the composite material. The pattern of heat flow is developed over the surface of a specimen after the application of a temperature gradient using some external source. This method is used to detect the flaws near to the surface of the test specimen by examining the pattern of heat flow through the surface. A temperature-sensitive coating is used to develop the heat flow pattern. Some commonly used thermography techniques are pulsed thermography, vibro thermography, active thermography, passive thermography and infrared thermography. Various authors carried out thermography-testing method to detect the presence of flaws in the structure of polymer composites. Chang et al. (2016) performed pulsed thermography testing of CF-reinforced polymer composite to identify the presence of flaws in the composite material. The authors concluded that thermographic data processing method removes the noise and non-uniform backgrounds, which makes it easier to identify the flaws. Fernandes et al. (2015) investigated the fibre orientation of CF-reinforced polymer composite using infrared thermography. The authors concluded that due to the randomness of the reinforced fibre with polymer matrix, point-by-point inspection was very time consuming. Therefore, the authors performed the flying laser-spot technique to detect the fibre orientation. Lisle et al. (2016) investigated the cracks of glass/epoxy composite using infrared thermography technique. Harizi et al. (2014) investigated the damage and internal flaws of glass/epoxy composite using passive infrared thermography. Meola et al. (2015) investigated the defect and impact damage of carbon/epoxy composite using infrared thermography.

6.10.1 INFRARED THERMOGRAPHY

In this method, the infrared cameras are used to detect the radiated heat. The cameras are used to detect the radiations and producing the images of those radiations that are called thermograms. All objects with a temperature above absolute zero emit infrared radiations. The variations in temperature affect the amount of radiation emitted and hence generating an image. This testing method is broadly classified in two categories: active thermography and passive thermography.

6.10.2 ACTIVE THERMOGRAPHY

A specimen that is in thermal equilibrium with the surrounding is brought into contact with a high-temperature component to produce thermal contrast. The surface-temperature variation of the object is analysed to detect the various flaws.

6.10.3 PASSIVE THERMOGRAPHY

This testing method is based on Ohm's law. An increase of electric resistance in the test specimen induces an increase of temperature. The temperature change is analysed to detect the defects. In this method, an electric current is supplied to the specimen to find out the variations of temperature. This technique is generally used for such components that are electrically conductive such as electrical appliances, switchboards and refrigerators.

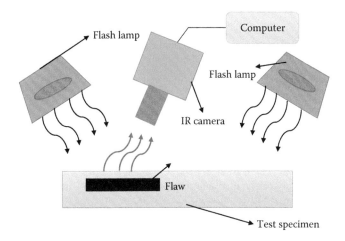

FIGURE 6.7   Working principle of pulsed thermography.

6.10.4 Pulsed Thermography

In this testing method, the composite specimen surface to be analysed is brought into contact with a hot or cold object, alternatively. A flash lamp is used with a coating of black paint as a heat source. An infrared camera is used to detect the contrast, which arises at the point of de-lamination or defects. The contrast depends on the thermal properties of composite material and the size of defects. Working principle of pulsed thermography is shown in Figure 6.7.

6.11   CONCLUDING REMARK

Inspection of polymer composites plays a very important role to assure safe working of the developed composite part. Flaws may develop during manufacturing of composites through various processing techniques and/or during machining of the composite. These defects weaken the fibre/matrix interfacial adhesion and lead to sudden failure of the composite component.

This chapter has discussed the working principles of various non-destructive testing methods, which are widely being used to evaluate the internal structure of polymer-based composite to identify the presence of defects within the composite surface and on the external surface. This chapter has also reviewed the work done by various researchers in this filed.

REFERENCES

Bergant, Z., J. Janez, and J. Grum. 2013. Ultrasonic testing of glass fiber reinforced composite with processing defects. *The 12th International Conference of the Slovenian Society for Non-Destructive Testing, Application of Contemporary Non-Destructive Testing in Engineering*, Portoroz, Slovenia: The Slovenian Society for Non-Destructive Testing, pp. 209–218.

Chang, Y.S., Z. Yan, K.H. Wang, and Y. Yao. 2016. Non-destructive testing of CFRP using pulsed thermography and multi-dimensional ensemble empirical mode decomposition. *Journal of the Taiwan Institute of Chemical Engineers*, 61: 54–63.

Cousins, R.R., and M.F. Markham. 1977. The use of ultrasonic spectroscopy in the location of delaminations in fibre-reinforced polymers. *Composites*, 8: 145–152.

Cuadra, J., P.A. Vanniamparambil, K. Hazeli, I. Bartoli, and A. Kontsos, 2013. Damage quantification in polymer composites using a hybrid NDT approach. *Composites Science and Technology*, 83: 11–21.

El-Sabbagh, A., L. Steuernagel, and G. Ziegmann. 2013. Ultrasonic testing of natural fibre polymer composites: effect of fibre content, humidity, stress on sound speed and comparison to glass fibre polymer composites. *Polymer Bulletin*, 70: 371–390.

Fernandes, H., H. Zhang, C.I. Castanedo, and X. Maldague. 2015. Fiber orientation assessment on randomly-oriented strand composites by means of infrared thermography. *Composites Science and Technology*, 121: 25–33.

Gholizadeh, S. 2016. A review of non-destructive testing methods of composite materials. *Procedia Structural Integrity*, 1: 50–57.

Gros, X.E. 1995. An eddy current approach to the detection of damage caused by low-energy impacts on carbon fibre reinforced materials. *Materials & Design*, 16: 167–173.

Harizi, W., S. Chaki, G. Bourse, and M. Ourak. 2014. Mechanical damage assessment of Glass Fiber-Reinforced Polymer composites using passive infrared thermography. *Composites: Part B*, 59: 74–79.

Icardi, U. 2003. Through-the-thickness displacements measurement in laminated composites using electronic speckle photography. *Mechanics of Materials*, 35: 35–51.

Imielinska, K., M. Castaings, R. Wojtyra, J. Haras, E.L. Clezio, and B. Hosten. 2004. Air-coupled ultrasonic C-scan technique in impact response testing of carbon fibre and hybrid: glass, carbon and Kevlar/epoxy composites. *Journal of Materials Processing Technology*, 157: 513–522.

Jesnitzer, F.E., and T. Winkler. 1981. Application of the holographic non-destructive testing method to the evaluation of disbondings in sandwich plates. *International Journal of Adhesion and Adhesives*, 1: 189–194.

Lisle, T., M.L. Pastor, C. Bouvet, and P. Margueres. 2016. Damage of woven composite under translaminar cracking tests using infrared thermography. *Composite Structures*, 161: 275–286. doi:10.1016/j.compstruct.2016.11.030.

Melin, L., J. Schon, and T. Nyman. 2002. Fatigue testing and buckling characteristics of impacted composite specimens. *International Journal of Fatigue*, 24: 263–272.

Meola, C., S. Boccardi, G.M. Carlomagno, N.D. Boffa, E. Monaco, and F. Ricci. 2015. Nondestructive evaluation of carbon fibre reinforced composites with infrared thermography and ultrasonics. *Composite Structures*, 134: 845–853.

Milwich, M. 2010. Types and production of textiles used for building and construction. In: Pohl G. (Ed.). *Textiles, Polymers and Composites for Buildings*, Cambridge, England: Woodhead Publishing, pp. 13–48.

Olsson, R., J. Iwarsson, L. Melin, A. Sjogren, and J. Solti. 2003. Experiments and analysis of laminates with artificial damage. *Composites Science and Technology*, 63: 199–209.

Orazioa, T.D., M. Leoa, A. Distantea, C. Guaragnellab, V. Pianesec, and G. Cavaccini. 2008. Automatic ultrasonic inspection for internal defect detection in composite materials. *NDT&E International*, 41: 145–154.

Ostwal, R.S., and A.V. Sawant. 2014. Ultrasonic testing of fiber reinforced polymer composites- An overview. *International Journal of Engineering Research & Technology (IJERT)*, 3: 544–547.

Pollock, A.A. 1969. Stress wave emission in ndt. *Non-Destructive Testing*, 2: 178–182.

Prade, F., F. Schaff, S. Senck, P. Meyer, J. Mohr, J. Kastner, and F. Pfeiffer. 2017. Nondestructive characterization of fiber orientation in short fiber reinforced polymer composites with X-ray vector radiography. *NDT&E International*, 86: 65–72.

Pye, C.J., and R.D. Adams. 1981. Detection of damage in fibre reinforced plastics using thermal fields generated during resonant vibration. *NDT International*, 14: 111–118.

Rathnakar, G., and P.P. Pandian. 2015. A review on the use and application of polymer composites in automotive industries. *International Journal for Research in Applied Science & Engineering Technology*, 3: 898–903.

Renoylds, W.N. 1985. Non-destructive testing (NDT) of fiber-reinforced composite materials. *Materials & Design*, 5: 256–270.

Reynolds, W.N., and S.J. Wilkinson. 1978. The analysis of fibre-reinforced porous composite materials by the measurement of ultrasonic wave velocities. *Ultrasonics*, 16: 159–163.

Schilling, P.J., B.P.R. Karedla, A.K. Tatiparthi, M.A. Verges, and P.D. Herrington. 2005. X-ray computed microtomography of internal damage in fiber reinforced polymer matrix composites. *Composites Science and Technology*, 65: 2071–2078.

Soutis, C. 2015. Introduction: Engineering requirements for aerospace composite materials. In: Phil E. Costas Soutis (Ed.). *Polymer Composites in the Aerospace Industry*, Cambridge, England: Woodhead Publishing, pp. 1–18.

Stone, D.E.W., and B. Clarke. 1975. Ultrasonic attenuation as a measure of void content in carbon-fibre reinforced plastics. *Non-destructive Testing*, 8: 137–145.

Sun, G., and Z. Zhou. 2016. Non-contact detection of delamination in layered anisotropic composite materials with ultrasonic waves generated and detected by lasers. *Optik*, 127: 6424–6433.

Unnporsson, R., M.T. Jonsson, and T.P. Runarsson. 2014. NDT methods for evaluating carbon fiber composites. *Rannis*, http://www.rannis.is.

Yakovlev, E.V., K.I. Zaytsev, I.N. Fokina, V.E. Karasik, and S.O. Yurchenko. 2014. Nondestructive testing of polymer composite materials using THz radiation. *Journal of Physics: Conference Series*, 486: 1–4.

Yashiro, S., J. Takatsubo, and N. Toyama. 2007. An NDT technique for composite structures using visualized Lamb-wave propagation. *Composites Science and Technology*, 67: 3202–3208.

7 Finite Element Analysis of Mechanical and Thermal Properties of Polymer Matrix Composites

Jasti Anurag, Sandhyarani Biswas and Siva Bhaskara Rao Devireddy

CONTENTS

7.1 INTRODUCTION

Nowadays, the interest in fibre-reinforced polymer composites (FRPCs) is increasing rapidly as these kinds of materials are used in the greatest diversity of applications due to its advantages such as high-specific strength and stiffness, design flexibility, excellent corrosion resistance, high durability and lightweight. In most of these applications, the properties of polymers are modified by using fibres as reinforcement to suit the high modulus/high-strength requirements. The properties of composite materials depend on various factors such as individual properties of the constituents, the micro-structural arrangement, the volume fractions of the constituents and the interfacial bonding between the reinforcement and matrix. The modelling and prediction of the properties of FRPCs based upon the knowledge of the individual properties of the constituents and their corresponding volume fractions are an active research area for the last few decades. A real understanding of their behaviour at various loading and environmental conditions (i.e. buckling, modal characteristics, deflections, stress, strain, temperature distribution, thermal conductivity, thermal diffusivity and specific heat capacity) is required. Prototyping is one of the commonly used techniques in testing before the actual production. It generally gives an overview of the functionality of the design and the necessary changes required to be made in the design. Apart from many advantages of prototyping, the main limitation is that it leads to implementation and then repairing or changing of the design. Prototyping also involves a lot of valuable time and money. Finite element analysis (FEA) overcomes these limitations by reducing the time and money involved in the design analysis process. The importance of FEA is that it eliminates the inevitability of physical testing while still maintaining the scope for design optimisation.

7.2   FINITE ELEMENT ANALYSIS OF POLYMER COMPOSITES

Generally, the analysis of composite materials can be examined at two distinct levels: macro-mechanical approach and micro-mechanical approach. In macro-mechanical approach, each layer of the composites is considered as a homogeneous, orthotropic and elastic continuum (Chawla 2012b). Based on the known properties of the individual layers, the macro-mechanical analysis involves investigation of the interaction of the individual layers of the laminate and their effect on the overall response of the laminate. Although the macro-mechanical analysis has the advantage of easiness, it is not capable of identifying the stress/strain states in the matrix and fibre level. In contrast, in the micro-mechanical approach, the fibre and matrix materials are individually considered to predict the overall response of the composite as well as the damage mechanisms and damage propagation in the composite (Agarwal and Broutman 1990; Daniel and Ishai 1994; Siuru and Busick 1994; Voyiadjis and Kattan 2005). The micro-mechanical analysis can be done to predict the effective properties of the composites from the knowledge of the individual constituents. The recent dramatic growth in computational capability for mathematical modelling and simulation increases the possibilities that the micro-mechanical methods can play an important role in the analysis of composite materials. To proceed with any modelling technique, it is first necessary to determine the mechanical properties of unidirectional composite at the appropriate fibre volume fraction. The representative volume element (RVE) or representative unit cell can be used in the micro-mechanics to calculate the effective properties of composite materials (Drugan and Willis 1996). In periodic composite structure, it is not necessary to study the whole structure, but only a unit cell is generally analysed for predicting the properties of the composite material. The basic unit of a unidirectional composite material is depicted in Figure 7.1. As a result of the unidirectional fibre alignment in the matrix, the resulting composite is strong and stiff along the fibre direction, whereas it is weak and soft in the other two transverse directions (Tenek and Argyris 2013). The axis longitudinal to the fibre direction is always taken as the primary axis, and the remaining two axes perpendicular to fibre direction are taken as the secondary axes.

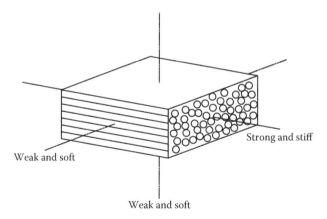

FIGURE 7.1 Basic unit of the unidirectional composite.

In FRPCs, the fibres are generally distributed randomly with different resin contents in different local areas, and the cross section of the fibres is not always uniform with the same dimensions. As a result, it may be difficult to predict the effective properties of composites by using micro-mechanical analysis. In order to simplify the analysis, few assumptions are generally considered to all micro-mechanical models: the fibres are having a uniform cross section, locally both the fibres and matrix are homogeneous and isotropic, both the fibre and matrix are linear elastic, and there is a perfect bonding between them. The composites are presumed to be free of irregularities and voids. Also, the geometry, distribution, alignment and the arrangement of fibres are considered to be regular. These assumptions significantly reduce the complexity of calculation and make it possible to perform the analysis of FRPCs with a RVE or unit cell.

7.2.1 Mechanical and Thermal Behaviours of Fibre-Reinforced Polymer Composites

The mechanical and thermal responses of composite materials are the challenging problem requiring expertise in a wide range of fields ranging from quantum mechanics to continuum mechanics depending on the length scale that is being studied. The FRPCs are generally transversely isotropic or orthotropic (Gardener 1994). This property makes it challenging for the analysts to examine their behaviour, which is necessary for the design process of any structure. The mechanical and thermal properties of polymer composites determine the need of it for a suitable application. These properties determine a product's functionality and durability over the expected product lifetime. The mechanical and thermal analyses are the preliminary steps in the process of designing a composite structure. A great deal of work has been done on the mechanical and thermal behaviours of FRPCs using different modelling techniques. The effective elastic modulus of composite materials by using the micro-mechanical modelling based on the generalised method of cells has been studied (Babu et al. 2005). The influence of micro-structural parameters such as the aspect ratio, shape and concentration of fillers on the elastic properties has also been discussed. A good overview on the micro-mechanical models that are used to predict the elastic moduli and Poisson's ratios of short fibre-reinforced composites has been presented (Tucker and Liang 1999). The accuracy of these models is evaluated by comparison with numerical finite element results. The study revealed that the Mori–Tanaka model gives the most accurate predictions followed by Halpin–Tsai equations. The micro-mechanical and thermal behaviours of glass FRPCs have been studied (Patnaik et al. 2012). The study revealed that the experimental results are in good agreement with the finite element model based on the representative area element approach. A micro-mechanical analysis of hybrid composites to evaluate the stiffness and strength properties using finite element method (FEM) has been carried out (Banerjee and Sankar 2014). The results were compared with the existing empirical and semi-empirical relations. The elastic properties of polypropylene fibre-reinforced composite material have been evaluated by using RVE model (Pal and Haseebuddin 2012). ANSYS software has been used to model the composite with different volume fractions and found that there is a good agreement between the finite element simulation results and the analytical results. The elastic modulus of the wood flour/kenaf fibre-reinforced polypropylene composites with a different weight ratio of the reinforcement has been studied (Mirbagheri et al. 2007). The investigation revealed that the rule of hybrid mixture shows a good agreement with the experimental values than those with the Halpin–Tsai equation. The elastic modulus of different types of natural fibre-reinforced high-density polyethylene composites has been studied by varying the volume fraction of the fibres using various modelling approaches (Facca et al. 2006).

The geometry and arrangement of fibre have a significant influence on the properties of the polymer composites. The effect of fibre geometry and RVE on the elastic properties of unidirectional glass fibre-reinforced epoxy composites has been studied (Devireddy and Biswas 2014b). In their study, two different types of fibre arrangement and two different types of fibre geometry have been considered. Figure 7.2a and b shows the arrangement of fibres in square RVE with circular fibres and square RVE with square fibres, respectively. Similarly, Figure 7.2c and d shows the arrangement of fibres in hexagonal RVE with circular fibres and hexagonal RVE with square fibres, respectively. Figure 7.3a–d shows the unit cells of the arrangements illustrated in Figure 7.2.

Both the analytical and numerical approaches have been used for analysing the elastic properties of the composites. In the numerical approach, FEA has been used and in analytical approach, periodic micro-structure, Halpin–Tsai model and the rule of mixture have been used. The effect of fibre volume fraction on the longitudinal modulus, transverse modulus, in-plane shear modulus and in-plane Poisson's

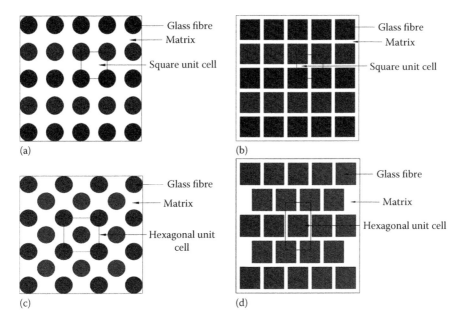

FIGURE 7.2 Arrangement of fibres in (a) square RVE with circular fibres, (b) square RVE with square fibres, (c) hexagonal RVE with circular fibres and (d) hexagonal RVE with square fibres. (Devireddy and Biswas 2014b.)

ratio of composites has been evaluated, and the results are shown in Figure 7.4a–d, respectively. The study revealed that the transverse modulus, longitudinal modulus and in-plane shear modulus of the composites increase and in-plane Poisson's ratio of the composites decreases with the increasing volume fraction of fibre. It has also been reported that the longitudinal modulus and the in-plane Poisson's ratio predicted by the FEA agree well with all the analytical predictions. However, the in-plane shear modulus and transverse modulus predicted by FEA with hexagonal RVE are closer to the periodic micro-structure model as compared to FEA with square RVE and other analytical methods.

Many semi-empirical and theoretical models have been suggested to determine the effective thermal conductivity of composites by the past researchers. A theoretical model of heat transfer in the polymer hollow micro-sphere composites based on the law of minimal thermal resistance and the equal law of the specific equivalent thermal conductivity has been developed (Liang and Li 2007). The effective transverse thermal conductivity of fibre-reinforced composites has been analysed by using FEM (Islam and Pramila 1999). Four different sets of thermal boundary conditions were applied to the square and circular cross section of fibres. The transverse thermal conductivity of continuous unidirectional fibre-reinforced composite materials using finite element has been studied (Grove 1990). An analytical model to

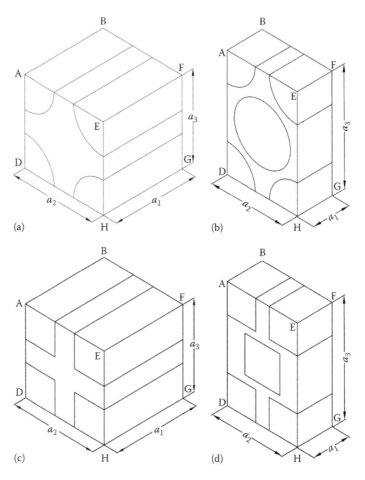

FIGURE 7.3   (a) Circular fibre with a square unit cell, (b) circular fibre with a hexagonal unit cell, (c) square fibre with a square unit cell and (d) square fibre with a hexagonal unit cell. (Devireddy and Biswas 2014b.)

determine the effective thermal conductivity of short banana fibre-reinforced epoxy composites has been studied (Sahu et al. 2014). The study revealed that the experimental values and the proposed model's values are in good agreement. A theoretical model for predicting the effective thermal conductivity of unidirectional fibre composite with square and hexagonal fibre geometry has been developed and observed that the results are in good agreement when the ratio of fibre and matrix is near to 1 (Xiao and Long 2014). The transverse thermal conductivity of laminated composites using both the analytical and numerical models has been studied and observed that the results are significantly influenced by the random fibre distributions, especially at higher fibre volume fractions (Sihn and Roy 2011). The thermal conductivity of short carbon FRPCs has been analysed by considering the effect of fibre orientation and fibre length (Fu and Mai 2003). The thermal conductivity of fibre-reinforced composites through the RVEs with single and multiple fibres has been estimated by using FEA based on homogenisation technique (Cao et al. 2012). The numerical and experimental analyses of thermal properties of date palm fibre-reinforced composites have been studied by using guarded-hot-plate method and three-dimensional (3D) finite element modelling (Haddadi et al. 2015). The in-plane and transverse thermal conductivities of hemp FRPCs for different volume fractions of fibre have been evaluated (Behzad and Sain 2007). The experimental results were compared with finite element model and two theoretical models and found a good agreement between the models and the obtained results.

The influence of fibre geometry and RVE on the thermal properties of unidirectional glass fibre/epoxy composites has been studied (Devireddy and Biswas 2014b). In their study, the longitudinal and transverse thermal conductivities have been predicted by the steady-state heat transfer analysis using FEA, and the results have been

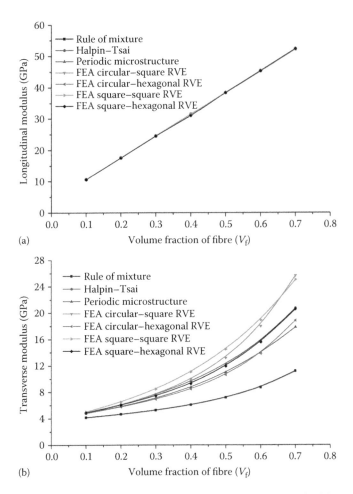

FIGURE 7.4   (a) Longitudinal modulus, (b) transverse modulus, (Devireddy and Biswas 2014b.)   *(Continued)*

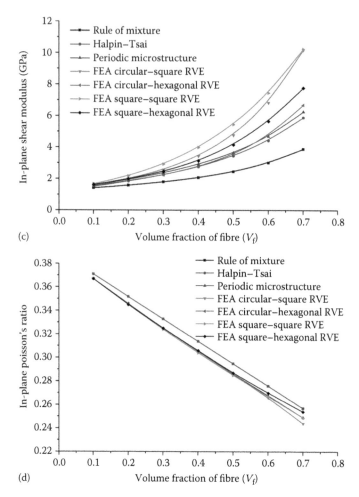

FIGURE 7.4 (Continued) (c) in-plane shear modulus and (d) in-plane Poisson's ratio with varying fibre volume fraction. (Devireddy and Biswas 2014b.)

compared with the existing analytical methods. Figure 7.5a and b shows the effect of fibre volume fraction on the longitudinal thermal conductivity and transverse thermal conductivity of the composites, respectively. The study revealed that the longitudinal thermal conductivity predicted by FEA agrees well with all analytical methods. However, the transverse thermal conductivity predicted by the FEA with hexagonal array is closer to the Hashin model as compared to rule of mixture and Chawla model.

The thermal behaviour of unidirectional banana/jute fibre-reinforced epoxy-based hybrid composites has been studied (Devireddy and Biswas 2016a). The schematic diagram of the composite where the fibres are arranged in the square array is shown in Figure 7.6. A micro-mechanical model has been developed to determine

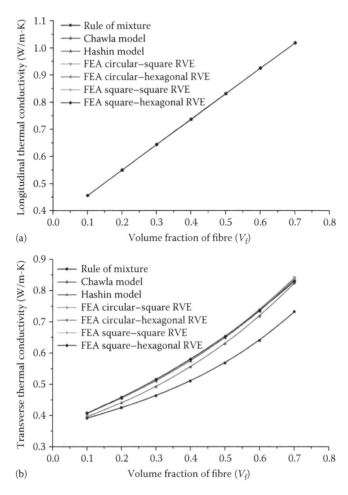

FIGURE 7.5 (a) Longitudinal thermal conductivity and (b) transverse thermal conductivity with varying fibre volume fraction. (Devireddy and Biswas 2014b.)

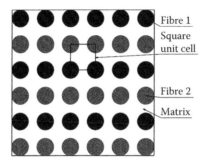

FIGURE 7.6 Unit cell of a hybrid composite. (Modified from Devireddy, S. B. R., and S. Biswas, *Proc. IME J. Mater. Des. Appl.*, 1–13, 2016.)

the transverse thermal conductivity using the law of minimal thermal resistance and equal law of specific equivalent thermal conductivity. The results have been validated with the results obtained by the experimental, FEA and existing analytical methods. The study revealed that the experimental and numerical results are in close approximation with the values predicted by the micro-mechanical model developed in their work. The study also reported that the longitudinal thermal conductivity of FEA results is in good agreement with the experimental results and the rule of hybrid mixture as compared to the Halpin–Tsai and Lewis and Nielsen models. The transverse thermal conductivity predicted by the proposed model and FEA is in good agreement with the experimental values.

To analyse the various properties of polymer composites using FEA, considering RVE method holds good for long fibre- as well as short fibre-reinforced composites (Berger et al. 2007; Pal and Haseebuddin 2012). Many researchers have used the RVE method for determining the properties of short fibre-based polymer composites. The thermal properties of short banana/jute fibre-reinforced epoxy-based hybrid composites have been studied (Devireddy and Biswas 2016b). Figure 7.7a shows the arrangement of short fibres in the composites. A single unit cell has been taken out from the square pattern of short fibre-based composites for their study and is shown in Figure 7.7b. A micro-mechanical model has been developed for evaluating the effective thermal conductivity of composites by using the law of minimal thermal resistance and the equal law of specific equivalent thermal conductivity. A numerical homogenisation technique based on the FEA has been used to evaluate the effective thermal conductivity. Finally, the results of the proposed model and FEA were validated with the results obtained from experimental and other analytical methods. It has been concluded from the study that the effective thermal conductivity predicted by the proposed model and FEA was in good agreement with the experimental values. It has also been reported that the specific heat capacity and thermal diffusivity of the composites decrease with an increase in the fibre loading.

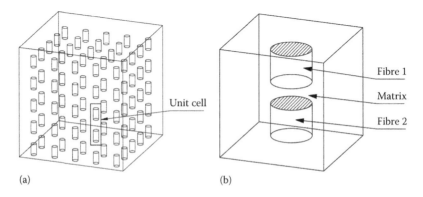

(a)   (b)

FIGURE 7.7   (a) Arrangement of short fibres in composites and (b) unit cell of hybrid fibre. (Modified from Devireddy, S. B. R., and S. Biswas, *J. Reinf. Plast. Compos.*, 35, 1157–1172, 2016.)

7.2.2   EFFECT OF FIBRE–MATRIX INTERPHASE ON THE PROPERTIES OF FIBRE-REINFORCED POLYMER COMPOSITES

In FRPCs, the main constituents are the reinforcement phase, the matrix phase and the interphase. Interphase is the region in which the material properties switch from matrix to fibre or vice versa (Chawla 2012a). In a composite material, transfer of load from the matrix to the fibre is done by fibre–matrix interphase. The inhomogeneous interphase has an effect on the material properties of fiber reinforced composite (FRCs) (Hwang and Gibson 1993; Jasiuk and Kouider 1993). FEA can be used to determine the effect of interphase on the various properties of composites through unit cell or RVE method. Figure 7.8a and b shows the unit cell of a composite with and without the interphase, respectively. Generally, it is assumed that each unit cell in the composite has same deformation mode and has no overlapping or separation between the neighbouring unit cells (Berger et al. 2007).

The effect of interphase on the performance of FRPCs has been studied by many researchers. The influence of interphase on the transverse Young's modulus of glass fibre-reinforced epoxy composites has been studied using a FEM model (Wacker et al. 1998). The average Young's modulus of the interphase has been identified by combining the theoretically calculated values with the experimental results. The study revealed that the material behaviour under mechanical load-oriented transverse to the fibre direction is influenced by the fibre treatment. The mechanical properties of unidirectional fibre-reinforced composite materials under transverse tensile loading at micro-mechanical level have been studied using FEA (Wang et al. 2011). The interphase has been represented by pre-inserted cohesive element layer between the fibre and matrix with tension and shear softening constitutive laws in order to evaluate the mechanical behaviours of composites. The storage moduli and internal damping properties of unidirectional fibre-reinforced composites including the interphase region using strain energy and a 2D finite element method have been studied (Hwang and Gibson 1993). Composite damping and moduli for transverse and longitudinal loading and in-plane and out of plane shear loading have been determined. The effect of the size of interphase on the composite damping has also been reported.

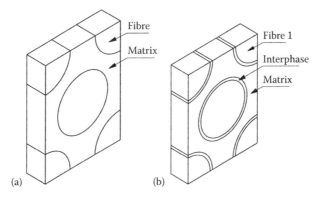

FIGURE 7.8   Unit cell of a composite (a) without interphase and (b) with interphase.

The influence of interphase on the mechanical behaviour of unidirectional Kevlar fibre/epoxy composites has been studied (Devireddy and Biswas 2014a). A 3D RVE model with a hexagonal packing geometry has been considered to predict the elastic properties by FEA at different fibre volume fractions. The analytical models such as periodic micro-structure, Halpin–Tsai model and the rule of mixture have also been used to validate the FEA results. It has been reported that the properties such as in-plane shear modulus, transverse modulus and longitudinal modulus of composite materials are increasing and the in-plane Poisson's ratio is decreasing with the increase in volume fraction of fibre. Figure 7.9a–d shows the influence of fibre volume fraction on the longitudinal modulus, transverse modulus, in-plane shear modulus and in-plane Poisson's ratio of composites, respectively. It has been concluded that the volume fraction of fibre and the interphase region significantly influences the elastic properties of composite materials.

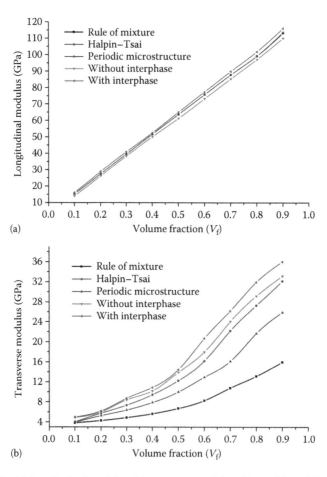

FIGURE 7.9 (a) Longitudinal modulus, (b) transverse modulus, (Devireddy and Biswas 2014a.)
(*Continued*)

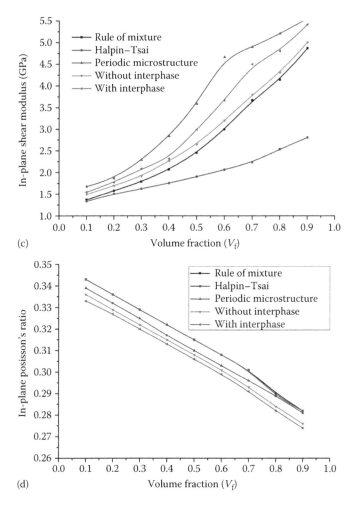

FIGURE 7.9 (Continued)   (c) in-plane shear modulus and (d) in-plane Poisson's ratio with varying fibre volume fraction. (Devireddy and Biswas 2014a.)

7.2.3 STATIC ANALYSIS OF COMPOSITE LAMINATE

A polymer composite lamina is made up of two main constituents: the fibre and the matrix. However, in real-life applications, composite laminates are used where laminates are a series of lamina or laminae stacked in a defined sequence to achieve the desirable material properties. Therefore, the knowledge of the mechanical behaviour of a lamina is essential for understanding the properties of laminated composite structure. Micro-mechanical modelling involves the investigation of the interaction of the individual layers of the laminate and their effect on the overall response of the laminate. The classical laminate theory is a widely accepted macro-mechanical approach for analysing the mechanical behaviour of composite laminates (Jones 1998). For a given stacking sequence, a composite's stress–strain relation can be

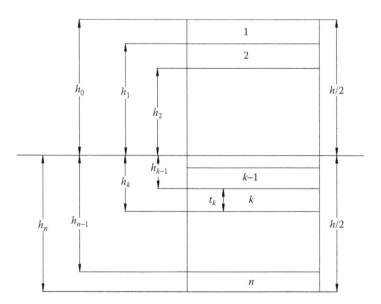

FIGURE 7.10 Location of composite lamina plies in a laminate.

determined, and the various coupling mechanisms between out-of-plane and in-plane deformation modes of a composite laminate can be explored. For laminates where the individual plies exhibit an inelastic response, additional terms are required in the classical laminate theory formulation, which incorporates inelastic strains into the formulation (Hidde and Herakovich 1992). FEA eliminates the tedious process of solving equations involved in the classical laminate theory. Different stresses may exist in the different lamina. Stresses in the individual lamina can be determined from static analysis using FEA (Pipes and Pagano 1974; Tahani and Nosier 2003a, 2003b; Nosier and Bahrami 2007; Sarvestani and Sarvestani 2011, 2012).

The location of a lamina in a composite laminate is depicted in Figure 7.10. The stresses in an individual lamina in a composite laminate can be determined using Equations 7.1 through 7.5 (Kaw 2005; Mukhopadhyay 2005). The static analysis of Kevlar fibre-reinforced epoxy composites with stacking sequences mentioned in Figure 7.11 has been studied using FEA. The properties of the unidirectional Kevlar fibre/epoxy composites with 50% of fibre loading have been taken from the previous work (Devireddy and Biswas 2014a). Figure 7.12 shows the stress in various layers or lamina in different composite laminates. From the graph, it is clear that the stress induced in a composite lamina depends on the stacking sequence and fibre orientation in the lamina.

$$
\begin{bmatrix} N_x \\ N_y \\ N_{xy} \\ M_x \\ M_y \\ M_{xy} \end{bmatrix} = \begin{bmatrix} A_{11} & A_{12} & A_{16} & B_{11} & B_{12} & B_{16} \\ A_{12} & A_{22} & A_{26} & B_{12} & B_{22} & B_{26} \\ A_{16} & A_{26} & A_{66} & B_{16} & B_{26} & B_{66} \\ B_{11} & B_{12} & B_{16} & D_{11} & D_{12} & D_{16} \\ B_{12} & B_{22} & B_{26} & D_{12} & D_{22} & D_{26} \\ B_{16} & B_{26} & B_{66} & D_{16} & D_{26} & D_{66} \end{bmatrix} \begin{bmatrix} \varepsilon_x^0 \\ \varepsilon_y^0 \\ \gamma_{xy}^0 \\ \kappa_x \\ \kappa_y \\ \kappa_{xy} \end{bmatrix} \tag{7.1}
$$

where:

N_x and N_y are the normal forces per unit length

M_x and M_y are the bending moments per unit length

N_{xy} is the shear force per unit length

M_{xy} is the twisting moment per unit length

$\begin{Bmatrix} \varepsilon_x^0 \\ \varepsilon_y^0 \\ \gamma_{xy}^0 \end{Bmatrix}$ are the mid-plane strains

$\begin{Bmatrix} \kappa_x \\ \kappa_y \\ \kappa_{xy} \end{Bmatrix}$ are the mid-plane curvatures

$$A_{ij} = \sum_{k=1}^{n} \left[\left(\bar{Q}_{ij} \right) \right]_k \left(h_k - h_{k-1} \right), \quad i = 1,2,6; \quad j = 1,2,6 \tag{7.2}$$

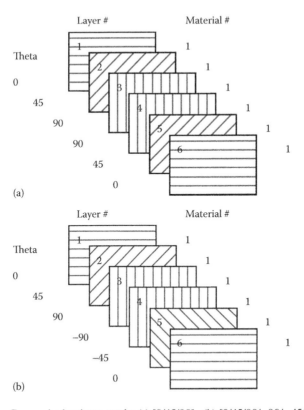

FIGURE 7.11 Composite laminate stacks (a) [0/45/90]$_s$, (b) [0/45/90/−90/−45/0],

(Continued)

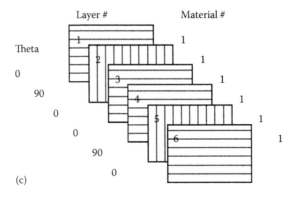

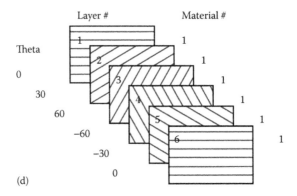

FIGURE 7.11 (Continued)   Composite laminate stacks (c) [0/90/0]$_s$ and (d) [0/30/60/−60/−30/0].

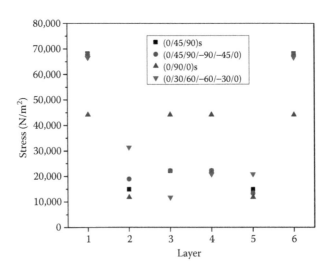

FIGURE 7.12   Stress in various layers of composite laminate.

$$B_{ij} = \frac{1}{2}\sum_{k=1}^{n}\left[\left(\bar{Q}_{ij}\right)\right]_k \left(h_k^2 - h_{k-1}^2\right), \quad i=1,2,6; \quad j=1,2,6 \qquad (7.3)$$

$$D_{ij} = \frac{1}{3}\sum_{k=1}^{n}\left[\left(\bar{Q}_{ij}\right)\right]_k \left(h_k^3 - h_{k-1}^3\right), \quad i=1,2,6; \quad j=1,2,6 \qquad (7.4)$$

The equation can be reduced to

$$\left\{\begin{matrix}\{N\}\\\{M\}\end{matrix}\right\} = \begin{bmatrix}[A] & [B]\\[B] & [D]\end{bmatrix}\left\{\begin{matrix}\{\varepsilon^0\}\\\{\kappa\}\end{matrix}\right\}$$

$$\Updownarrow \qquad\qquad (7.5)$$

$$\left\{\begin{matrix}\{\varepsilon^0\}\\\{\kappa\}\end{matrix}\right\} = \begin{bmatrix}[a] & [b]\\[b] & [d]\end{bmatrix}\left\{\begin{matrix}\{N\}\\\{M\}\end{matrix}\right\}$$

where:
  [A] is the in-plane stiffness matrix
  [a] is the in-plane compliance matrix
  [B] is the coupling stiffness matrix
  [b] is the coupling compliance matrix
  [D] is the bending stiffness matrix
  [d] is the flexural compliance matrix

7.2.4 FREE VIBRATION ANALYSIS OF COMPOSITE LAMINATE

FRPCs are used as the structural components where they are exposed to dynamic load-ing conditions. These components need to resist very harsh environment situations where the damage is mainly caused due to the resonant vibrations. Designing resonant-free structures is necessary as the FRPCs are used in numerous applications such as floors, runways and so on (Lal et al. 2008). Frequency and stress are generally used to determine the fracture vulnerability of the material due to the vibration. Vibrations can also be used to estimate the existence, size and location of delamination and also used as a tool for condition monitoring (Lee 2000; Kumar et al. 2014). Therefore, vibration analysis of the FRPCs is critical for the design of a structure.

Vibration analysis poses a great difficulty in the structural design of polymer composites. The Ritz and Rayleigh methods are the most common methods used for vibration analysis (Bhat 1985; Deobald and Gibson 1988; Trefethen and Bau III 1997; Gibson 2000). Over the years, many other methods have also been devel-oped by researchers (Singh et al. 2001; Hajianmaleki and Qatu 2013). Generally, the governing differential equations and their boundary conditions are mainly used for performing the vibration analysis. However, for real-life situations, the use of differential equations for vibration analysis of complicated composite structures is very difficult. To overcome this difficulty, FEA is mainly used. The free vibration analysis of unidirectional glass fibre-reinforced epoxy composites with stacking sequences mentioned earlier in Figure 7.11b and d has been studied using FEA. The properties of the unidirectional Kevlar fibre/epoxy composites with 50% of fibre loading have been taken from the previous work (Devireddy and Biswas 2014a). Six numbers of modes and the frequencies are determined as depicted in Figure 7.13. Figure 7.14 shows the mode shapes for different modes.

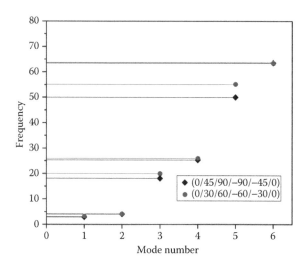

FIGURE 7.13 Natural frequencies for various modes.

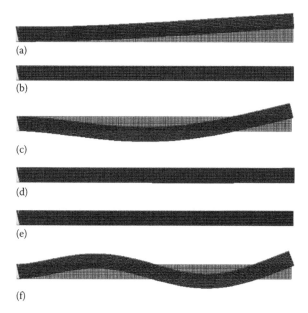

FIGURE 7.14 Mode shapes of stack [0/45/90/−90/−45/0] for mode (a) 1, (b) 2, (c) 3, (d) 4, (e) 5 and (f) 6.

7.3   CONCLUSION

Designing composite structures poses a great difficulty due to uncertainties in composite material properties. Therefore, a concurrent design is necessary. Many researchers have developed analytical methods to analyse the material properties of composites. However, due to the complexity, it is very difficult to analyse composites using the analytical methods. FEA proves to be an efficient tool in design and analysis of composite materials. The results obtained by FEA are found to be in good agreement with the experimental results of analytical methods. FEA reduces the time and saves money by eliminating the use of prototype for testing real-time behaviour of FRPCs.

REFERENCES

Agarwal, B. D., and L. J. Broutman. 1990. *Analysis and Performance of Fiber Composites.* SPE Monographs. Vol. 18. New York: Wiley.

Babu, P. E. J., S. Savithri, U. T. S. Pillai, and B. C. Pai. 2005. Micromechanical modeling of hybrid composites. *Polymer* 46 (18): 7478–7484.

Banerjee, S., and B. V. Sankar. 2014. Mechanical properties of hybrid composites using finite element method based micromechanics. *Composites Part B: Engineering* 58: 318–327.

Behzad, T., and M. Sain. 2007. Measurement and prediction of thermal conductivity for hemp fiber reinforced composites. *Polymer Engineering and Science* 47 (7): 977–983.

Berger, H., S. Kari, U. Gabbert, R. Rodríguez Ramos, J. Bravo Castillero, and R. Guinovart Díaz. 2007. Evaluation of effective material properties of randomly distributed short cylindrical fiber composites using a numerical homogenization technique. *Journal of Mechanics of Materials and Structures* 2 (8): 1561–1570.

Bhat, R. B. 1985. Natural frequencies of rectangular plates using characteristic orthogonal polynomials in Rayleigh-Ritz method. *Journal of Sound and Vibration* 102 (4): 493–499.

Cao, C., A. Yu, and Q.-H. Qin. 2012. Evaluation of effective thermal conductivity of fiber-reinforced composites. *International Journal of Architecture, Engineering and Construction* 1 (1): 14–29.

Chawla, K. K. 2012a. Interfaces. In *Composite Materials*, pp. 105–133. New York: Springer.

Chawla, K. K. 2012b. Macromechanics of composites. In *Composite Materials*, pp. 387–419. New York: Springer.

Daniel, I. M., and O. Ishai. 1994. *Engineering Mechanics of Composite Materials.* 2nd ed. New York: Oxford University Press.

Deobald, L. R., and R. F. Gibson. 1988. Determination of elastic constants of orthotropic plates by a modal analysis/Rayleigh-Ritz technique. *Journal of Sound and Vibration* 124 (2): 269–283.

Devireddy, S. B. R., and S. Biswas. 2014a. Micromechanical analysis of effect of interphase on mechanical properties of Kevlar fiber reinforced epoxy composites. *International Journal of Current Engineering and Technology* 2 (2): 115–120.

Devireddy, S. B. R., and S. Biswas. 2014b. Effect of fiber geometry and representative volume element on elastic and thermal properties of unidirectional fiber-reinforced composites. *Journal of Composites* 2014: 1–12.

Devireddy, S. B. R., and S. Biswas. 2016a. Physical and thermal properties of unidirectional Banana-Jute hybrid fiber-reinforced epoxy composites. *Journal of Reinforced Plastics and Composites* 35 (15): 1157–1172.

Devireddy, S. B. R., and S. Biswas. 2016b. Thermo-physical properties of short banana-jute fiber-reinforced epoxy-based hybrid composites. *Proceedings of the Institution of Mechanical Engineers, Part L: Journal of Materials: Design and Applications*, 1–13.

Drugan, W. J., and J. R. Willis. 1996. A micromechanics-based nonlocal constitutive equation and estimates of representative volume element size for elastic composites. *Journal of the Mechanics and Physics of Solids* 44 (4): 497–524.

Facca, A. G., M. T. Kortschot, and N. Yan. 2006. Predicting the elastic modulus of natural fibre reinforced thermoplastics. *Composites Part A: Applied Science and Manufacturing* 37 (10): 1660–1671.

Fu, S. Y., and Y. W. Mai. 2003. Thermal conductivity of misaligned short-fiber-reinforced polymer composites. *Journal of Applied Polymer Science* 88 (6): 1497–1505.

Gardener, J. P. 1994. Micromechanical modeling of composite materials in finite element analysis using an embedded cell approach. Doctoral dissertation, Massachusetts Institute of Technology.

Gibson, R. F. 2000. Modal vibration response measurements for characterization of composite materials and structures. *Composites Science and Technology* 60 (15): 2769–2780.

Grove, S. M. 1990. A model of transverse thermal conductivity in unidirectional fibre-reinforced composites. *Composites Science and Technology* 38 (3): 199–209.

Haddadi, M., B. Agoudjil, N. Benmansour, A. Boudenne, and B. Garnier. 2015. Experimental and modeling study of effective thermal conductivity of polymer filled with date palm fibers. *Polymer Composites*, pp. 1–8.

Hajianmaleki, M., and M. S. Qatu. 2013. Vibrations of straight and curved composite beams: A review. *Composite Structures* 100: 218–232.

Hidde, J. S., and C. T. Herakovich. 1992. Inelastic response of hybrid composite laminates. *Journal of Composite Materials* 26 (1): 2–19.

Hwang, S. J., and R. F. Gibson. 1993. Prediction of fiber-matrix interphase effects on damping of composites using a micromechanical strain energy/finite element approach. *Composites Engineering* 3 (10): 975–984.

Islam, M. R., and A. Pramila. 1999. Thermal conductivity of fiber reinforced composites by the FEM. *Journal of Composite Materials* 33 (18): 1699–1715.

Jasiuk, I., and M. W. Kouider. 1993. The effect of an inhomogeneous interphase on the elastic constants of transversely isotropic composites. *Mechanics of Materials* 15 (1): 53–63.

Jones, R. M. 1998. *Mechanics of Composite Materials*. Boca Raton, FL: CRC press.

Kaw, A. K. 2005. *Mechanics of Composite Materials, Second Edition*. 1st ed. Hoboken: CRC Press.

Kumar, T. P., A. Jasti, M. Saimurugan, and K. I. Ramachandran. 2014. Vibration based fault diagnosis of automobile gearbox using soft computing techniques. In *Proceedings of the 2014 International Conference on Interdisciplinary Advances in Applied Computing - ICONIAAC'14*, pp. 1–7. New York: ACM.

Lal, A., B. N. Singh, and R. Kumar. 2008. Nonlinear free vibration of laminated composite plates on elastic foundation with random system properties. *International Journal of Mechanical Sciences* 50 (7): 1203–1212.

Lee, J. 2000. Free vibration analysis of delaminated composite beams. *Computers & Structures* 74 (2): 121–129.

Liang, J. Z., and F. H. Li. 2007. Heat transfer in polymer composites filled with inorganic hollow micro-spheres: A theoretical model. *Polymer Testing* 26 (8): 1025–1030.

Mirbagheri, J., M. Tajvidi, J. C. Hermanson, and I. Ghasemi. 2007. Tensile properties of wood flour/kenaf fiber polypropylene hybrid composites. *Journal of Applied Polymer Science* 105 (5): 3054–3059.

Mukhopadhyay, M. 2005. *Mechanics of Composite Materials and Structures*. Hyderabad, India: Universities Press.

Nosier, A., and A. Bahrami. 2007. Interlaminar stresses in antisymmetric angle-ply laminates. *Composite Structures* 78 (1): 18–33.

Pal, B., and M. R. Haseebuddin. 2012. Analytical estimation of elastic properties of polypropylene fiber matrix composite by finite element analysis. *Advances in Materials Physics and Chemistry* 2 (1): 23–30.

Patnaik, A., P. Kumar, S. Biswas, and M. Kumar. 2012. Investigations on micro-mechanical and thermal characteristics of glass fiber reinforced epoxy based binary composite structure using finite element method. *Computational Materials Science* 62: 142–151.

Pipes, R. B., and N. J. Pagano. 1974. Interlaminar stresses in composite laminates-an approximate elasticity solution. *Transactions of the ASME* 41: 668–672.

Sahu, Y. K., J. Banjare, A. Agrawal, and A. Satapathy. 2014. Establishment of an analytical model to predict effective thermal conductivity of fiber reinforced polymer composites. *International Journal of Plastics Technology* 18 (3): 368–373.

Sarvestani, H. Y., and M. Y. Sarvestani. 2011. Interlaminar stress analysis of general composite laminates. *International Journal of Mechanical Sciences* 53 (11): 958–967.

Sarvestani, H. Y., and M. Y. Sarvestani. 2012. Free-edge stress analysis of general composite laminates under extension, torsion and bending. *Applied Mathematical Modelling* 36 (4): 1570–1588.

Sihn, S., and A. K. Roy. 2011. Micromechanical analysis for transverse thermal conductivity of composites. *Journal of Composite Materials* 45 (11): 1245–1255.

Singh, B. N., D. Yadav, and N. G. R. Iyengar. 2001. Natural frequencies of composite plates with random material properties using higher-order shear deformation theory. *International Journal of Mechanical Sciences* 43 (10): 2193–2214.

Siuru, W. D., and J. D. Busick. 1994. *Future Flight: The Next Generation of Aircraft Technology*. New York: McGraw-Hill Professional.

Tahani, M., and A. Nosier. 2003a. Three-dimensional interlaminar stress analysis at free edges of general cross-ply composite laminates. *Materials & Design* 24 (2): 121–130.

Tahani, M., and A. Nosier. 2003b. Free edge stress analysis of general cross-ply composite laminates under extension and thermal loading. *Composite Structures* 60 (1): 91–103.

Tenek, L. T., and J. Argyris. 2013. *Finite Element Analysis for Composite Structures*. Vol. 59. Solid Mechanics and Its Applications. Dordrecht, the Netherlands: Springer.

Trefethen, L. N, and D. Bau III. 1997. *Numerical Linear Algebra*. Philadelphia, PA: Society for Industrial and Applied Mathematics.

Tucker, C. L., and E. Liang. 1999. Stiffness predictions for unidirectional short-fiber composites: Review and evaluation. *Composites Science and Technology* 59 (5): 655–671.

Voyiadjis, G. Z., and P. I. Kattan. 2005. *Mechanics of Composite Materials with MATLAB. Mechanics of Composite Materials with MATLAB*. Berlin, Germany: Springer-Verlag.

Wacker, G., A. K. Bledzki, and A. Chateb. 1998. Effect of interphase on the transverse young' S modulus of glass/epoxy composites. *Composites Part A: Applied Science and Manufacturing* 29 (5–6): 619–626.

Wang, R. M., S. R. Zheng, and Y. G. Zheng. 2011. *Polymer Matrix Composites and Technology*. Cambridge: Woodhead Publishing.

Xiao, X., and A. Long. 2014. A solution for transverse thermal conductivity of composites with quadratic or hexagonal unidirectional fibres. *Science and Engineering of Composite Materials* 21 (1): 99–109.

8 Tribo-Performance Evaluation and Optimisation of Fibre-Reinforced Phenolic-Based Friction Composites

Tej Singh, Amar Patnaik and Ranchan Chauhan

CONTENTS

8.1 INTRODUCTION

Brake system is one of the important parts of the automobiles, locomotives, aircrafts and other moving bodies because it is related to the safety of human life and machines [1]. Automotive braking system is standout, the most imperative utilisations of polymer-based friction composite materials [2]. Polymer-based composite friction materials is the most intricate composite material as it contains a polymer matrix inside which various ingredients are circulated to accomplish the coveted stringent level of braking goals such as steady and high coefficient of friction (COF), low fade, high recovery, low wear and low noise under differing working paces and operating conditions [2,3]. The friction materials designer suggested that such material must have an amalgam of multi-ingredients that are broadly classified under the binder resin, reinforcing fibres, fillers and property modifiers [4]. The selection of appropriate ingredients, understanding the braking interface tribology and developing the friction composite materials that are fulfilling aforesaid norms are the real challenge faced by the formulation designers [5–10].

Among the many ingredients, fibres such as organic, inorganic, ceramic and their various combinations have been found to play a crucial role to enhance the tribo performance of brake friction composites [11,12]. Various metallic fibres such as aluminium fibre, steel fibre, copper and brass fibres are extensively studied as fibrous reinforcement in friction composite materials [13–15]. Among all the metallic fibres, copper fibre stands extremely important because its high thermal conductivity. The inclusion of copper fibre in friction composites not only results in reduced tribo-interface temperature, but also leads to stable and high COF with reduced wear [16, 17]. It is true that copper remains almost as inevitable ingredient of friction formulations, but recently it is reported as a threat to the aquatic life [18]. Brass, an alloy of copper was also reported to display good tribological properties. According to the literature, the main impact of brass is the improvement of thermal properties of brake friction materials, which reduces the accumulation of frictional heat at the contact interface and also limits the degradation of the organic components of the friction material, which results in improved performance [19,20]. Like brass, bronze is also an alloy of copper and possesses good thermal conductivity. The influence of bronze in friction formulation is hardly reported in literature. So the influence of bronze on the performance of brake friction formulations can be studied.

The friction composite materials for braking application require an appropriate set of performance criteria, such as negligible fading, faster recovery, better wear resistance and low sensitivity towards load-speed alterations, for reliable and safe performance [1]. The evaluation of such criteria and optimal selection of desired formulation have multi-level and multi-factor features, so such difficulties can be regarded as multiple criteria decision-making (MCDM) [2–6]. The MCDM methods are the prospective quantitative approaches for solving the decisive problems involving finite number of alternatives and criteria. In the past, several researchers used different MCDM approaches such as preference ranking organisation method for enrichment evaluations (PROMETHEE) [21], grey relation analysis (GRA) [10], technique for order preference by similarity to ideal solution (TOPSIS) [6], elimination and choice translating reality (ELECTRE) [22], preference selection index (PSI) [5,23], vise kriterijumska optimizacija kompromisno resenje (VIKOR) [8] and analytical hierarchy process (AHP) [6] in various areas such as management, engineering and science. Among them, PROMETHEE is popular and quite simple ranking method used to rank a finite set of alternatives [21].

In the light of earlier literature, the target of this research work is to study the influence of bronze fibre addition on the tribological properties and to find the most desirable friction composites by PROMETHEE II method.

8.2 MATERIALS AND METHODOLOGY

8.2.1 FABRICATION OF FRICTION COMPOSITES

Friction composites based on parent composition (50 wt.%) and varying proportions of bronze fibres to barium sulphate (50 wt.%) were designed accordingly as per Table 8.1. The details of the methodology adopted for composite fabrication are briefly reported elsewhere [24].

8.2.2 TRIBO-PERFORMANCE EVALUATION METHODOLOGY

Tribological properties of the fabricated brake pads were measured on a Krauss-type friction tester. The schematic of Krauss tester and detailed working of the machine are reported elsewhere [25]. The standard regulatory test procedure, pulse velocity wave-3212 (PVW-3212), confirming to protocol R-90 of ECE R-90 has been adopted for the evaluation of tribological characteristics of the fabricated brake pads [26].

TABLE 8.1
Composition and Nomenclature Details of Prepared Friction Composites

Composition (wt.%)	Composite Designation		
	FB-1	FB-2	FB-3
Parent composition[a]	50	50	50
Barium sulphate	45	40	35
Bronze fibre	5	10	15

[a] Parent composition: Phenolic resin – 15 wt.%; fibres (Kevlar, Lapinus) – 15 wt.%; Graphite – 5 wt.%, Alumina – 5 wt.%, Potassium titanate – 5 wt.% and Vermiculite – 5 wt.%. FB = Friction composites based on bronze fibre.

8.2.3 Optimisation of Tribo-Performance

In this work, PROMETHEE II approach is used to rank the friction composites. The various steps of the proposed algorithm can be expressed as follows:

Step I: Determination of alternatives and criteria
In this phase, the number of alternatives $(A_i, i = 1,2...m)$ and criteria $(C_j, j = 1,2...n)$ for a given MCDM problems are identified.
Step II: Construction of decision matrix
After the identification of alternatives and criteria, the value of experimental results for each alternative with respected criterion is expressed in the form of a decision matrix as

$$D = \begin{array}{c} \\ A_1 \\ A_2 \\ \vdots \\ A_i \\ \vdots \\ A_m \end{array} \begin{array}{ccccc} C_1 & C_2 & \cdots & C_j & \cdots & C_n \\ \left[\begin{array}{ccccc} d_{11} & d_{12} & \cdots & d_{1j} & \cdots & d_{1n} \\ d_{21} & d_{22} & \cdots & d_{2j} & \cdots & d_{2n} \\ \vdots & \vdots & \cdots & \vdots & \cdots & \vdots \\ d_{i1} & d_{i2} & \cdots & d_{ij} & \cdots & d_{in} \\ \vdots & \vdots & \cdots & \vdots & \cdots & \vdots \\ d_{m1} & d_{m2} & \cdots & d_{mj} & \cdots & d_{mn} \end{array}\right] \end{array} \tag{8.1}$$

where, element d_{ij} represents the value of *j*th criterion (C_j) for the *i*th alternative (A_i).
Step III: Selection of optimal composition: PROMETHEE II method
First of all, the decision matrix is normalised by using the following equations:
Larger the better,

$$d'_{ij} = \frac{d_{ij} - \min\{d_{ij}\}}{\max\{d_{ij}\} - \min\{d_{ij}\}} \tag{8.2}$$

Smaller the better,

$$d'_{ij} = \frac{\max\{d_{ij}\} - d_{ij}}{\max\{d_{ij}\} - \min\{d_{ij}\}} \tag{8.3}$$

In the next step, the preference functions $(\partial_j(i,i'))$ are calculated for each criterion as

$$\partial_j(i,i') = 0 \quad \text{if } d'_{ij} \le d'_{i'j}, \text{ and } \partial_j(i,i') = (d'_{ij} - d'_{i'j}) \quad \text{if } d'_{ij} > d'_{i'j} \tag{8.4}$$

Thereafter, the weighted preference function is determined by using the following equation:

$$\varpi(i,i') = \sum_{j=1}^{n} \omega_j \times \partial_j(i,i') \tag{8.5}$$

where ω_j is the weight of the selected criteria and is calculated using entropy method [7].

In the next step, the positive and negative outranking flows are determined by using the following equations:

$$\hbar_i^+ = \frac{1}{m-1} \sum_{i'=1}^{m} \varpi(i,i') \quad \text{and} \quad \hbar_i^- = \frac{1}{m-1} \sum_{i'=1}^{m} \varpi(i',i) \tag{8.6}$$

Finally the net outranking flow for each alternative is calculated as follows:

$$\lambda_i = \hbar_i^+ - \hbar_i^- \tag{8.7}$$

Thus, the best alternative is the one having the highest λ_i value.

8.3 RESULTS AND DISCUSSION

8.3.1 COEFFICIENT OF FRICTION IN RELATION TO COMPOSITION

The COF is the average COF taken after 1 s for fade and recovery cycles at temperature greater than 100°C. It is observed that COF remains in the range of 0.391–0.429 and found to decrease with increase in bronze fibre as shown in Figure 8.1. Decreased COF may be attributed due to the uniformity of fibrous ingredients in the

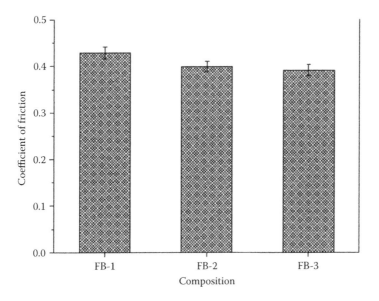

FIGURE 8.1 Coefficient of friction versus composition.

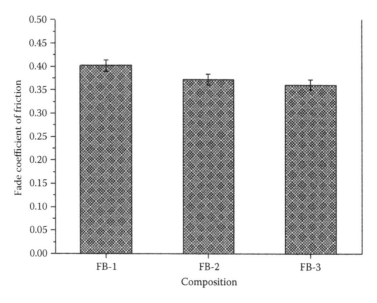

FIGURE 8.2   Fade coefficient of friction versus composition.

phenolic matrix and the ability to form structurally stable friction films, which help
in the reduction of friction performance at elevated temperatures.

8.3.2   FADE COEFFICIENT OF FRICTION IN RELATION TO COMPOSITION

Fade coefficient of friction (fade-COF) is the minimum COF for the fade cycles
taken after 270°C temperature and is shown in Figure 8.2. It is observed that Fade-
COF remains in the range of 0.361–0.402 and found to decrease with increase in
bronze fibre. This decrease in fade-COF may be attributed to the promotions of
the friction film formation, which leads to increase in the contact area of brake pad
and disc. This increased contact area reduced the applied load on the brake pad and
resulted in fade.

8.3.3   RECOVERY COEFFICIENT OF FRICTION IN RELATION TO COMPOSITION

Recovery coefficient of friction (recovery-COF) is the maximum COF for the
recovery cycle taken after 100°C temperature. It is observed that recovery-COF
(shown in Figure 8.3) remains in the range of 0.392–0.431. Bronze fibre inclusion
appeared to be detrimental from recovery-COF point of view. The higher the
fibre content, the less was the recovery-COF. The recovery-COF remains higher
than that of COF and fade-COF; this increase may be due to disintegration of
loosely compacted wear debris during the recovery cycle leading to the formation of third bodies between the pad and disc interface. These hardened third
bodies destroy the friction films formed during cold and fade cycles leading to
an increase in friction.

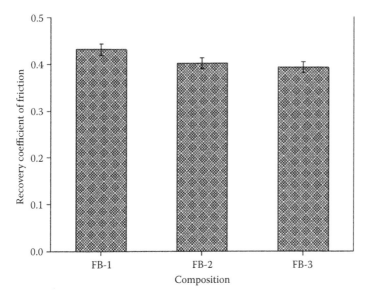

FIGURE 8.3 Recovery coefficient of friction versus composition.

8.3.4 WEAR PERFORMANCE IN RELATION TO COMPOSITION

Wear performance of the friction composites is measured in terms of weight loss (in grams) of the composite before and after tribological characterisation. The wear performance of the investigated composites followed the trend: FB-3 > FB-2 > FB-1 (Figure 8.4). Figure 8.4 revealed that wear performance is higher for the composites

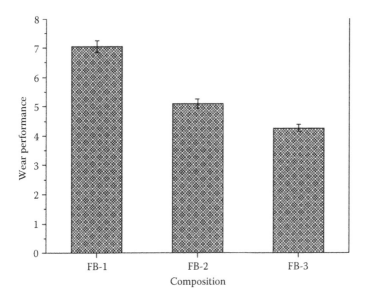

FIGURE 8.4 Wear performance of friction versus composition.

containing higher bronze fibre than that obtained from the composites having lower bronze fibre content. The lower value of wear may be due to the formation of strong friction film that sticks nicely with the brake pad and disc surfaces, which help in minimising wear and friction.

It is clearly observed from Figures 8.1 through 8.4 that no single composition showed the best tribo performance when all the criteria were taken into account simultaneously. Hence, it is a difficult task to suggest a composite for which highest tribo performance will be achieved. So, entropy-PROMETHEE II methodology is therefore carried out to come out with the optimal composition, which delivers the highest tribo performance.

8.3.5 Ranking of the Friction Composites

The experimental data of the friction composites in terms of four criteria, namely COF, fade-COF, recovery-COF and wear are reported in Table 8.2, which involves profit as well as loss criteria. Among these criteria, COF, fade-COF and recovery-COF have been treated as profit, and wear has been treated as loss criterion.

The problem has been solved by PROMETHEE II method considering criterion weight as determined by the entropy method. For PROMETHEE II method, first the decision matrix as shown in Table 8.2 in the form of Equation 8.1 was normalised in the range of 0–1 by using Equations 8.2 and 8.3 and is given in Table 8.3. After normalisation, the preference functions and weighted preference functions for all pairs of alternatives were determined by using Equations 8.4 and 8.5 and are listed in Tables 8.4 and 8.5. After that, positive outranking, negative outranking and net outranking, flows of the alternatives were obtained by using Equations 8.6 and 8.7 and are listed in Table 8.6. The final ranking of friction composite was obtained as FB-3 > FB-2 > FB-1. It is observed that composite

TABLE 8.2
Experimental Data Used as Decision Matrix

Alternative	COF	Fade-COF	Recovery-COF	Wear
FB-1	0.429	0.402	0.431	7.05
FB-2	0.399	0.372	0.401	5.10
FB-3	0.391	0.361	0.392	4.25

TABLE 8.3
Normalised Decision Matrix

Alternative	COF	Fade-COF	Recovery-COF	Wear
FB-1	1.0000	0.0000	1.0000	0.0000
FB-2	0.2105	0.7317	0.2308	0.6964
FB-3	0.0000	1.0000	0.0000	1.0000
Weight	0.0336	0.0429	0.0345	0.8890

TABLE 8.4
Preference Functions of the Alternatives

Alternative	COF	Fade-COF	Recovery-COF	Wear
(FB-1, FB-2)	0.7895	0.0000	0.7692	0.0000
(FB-1, FB-3)	1.0000	0.0000	1.0000	0.0000
(FB-2, FB-1)	0.0000	0.7317	0.0000	0.6964
(FB-2, FB-3)	0.2105	0.0000	0.2308	0.0000
(FB-3, FB-1)	0.0000	1.0000	0.0000	1.0000
(FB-3, FB-2)	0.0000	0.2683	0.0000	0.3036

TABLE 8.5
Weighted Preference Functions of the Alternatives

	FB-1	FB-2	FB-3
FB-1		0.0531	0.0681
FB-2	0.6505		0.0150
FB-3	0.9319	0.2814	

TABLE 8.6
Positive, Negative and Net Outranking Flows with Ranking of the Alternatives

Alternatives	$\hbar_i^+$	$\hbar_i^-$	λ_i	Ranking
FB-1	0.0606	0.7912	−0.7306	3
FB-2	0.3328	0.1672	0.1656	2
FB-3	0.6066	0.0416	0.5650	1

FB-3 indicates the most preferred friction performances with net outranking flow of 0.5650 and hence exhibits the optimal performance.

8.4 CONCLUSIONS

The performance assessment of tribological properties of bronze fibre-reinforced phenolic base friction composites has been carried out in this work. Varying barium sulphate and bronze fibre proportions in the formulation mix have resulted in stable COF, better wear resistance and recovery characteristics indicating the synergism between these ingredients. Also, the increased bronze fibre content in the formulation mix causes a decrease in COF and fade-COF, respectively. By applying

PROMETHEE II approach, the final ranking of polymer-based friction composite was obtained as FB-3 > FB-2 > FB-1, and the composition FB-3 with 15 wt.% bronze fibre exhibits the optimal tribological properties.

REFERENCES

1. Fred P. *Brake Handbook*. Tucson, AZ: HP Books, 1985.
2. Bijwe J. Composites as friction materials: Recent developments in non-asbestos fiber reinforced friction materials-A review. *Polymer Composites* 1997; 18:378–396.
3. Chan D. and Stachowiak G.W. Review of automotive brake friction materials. *Proceedings of the Institution of Mechanical Engineers Part D: Journal of Automobile Engineering* 2004; 218(9): 953–966.
4. Singh T. Tribo-performance evaluation of fibre reinforced and nano-filled composite friction materials. PhD thesis, NIT Hamirpur, 2013.
5. Singh T., Patnaik A., Gangil B., and Chauhan R. Optimization of tribo-performance of brake friction materials: Effect of nano filler. *Wear* 2015; 324–325: 10–16.
6. Singh T., Patnaik A., and Satapathy B.K. Development and optimization of hybrid friction materials consisting of nanoclay and carbon nanotubes by using analytical hierarchy process (AHP) and technique for order preference by similarity to ideal solution (TOPSIS) under fuzzy atmosphere. *Walailak Journal of Science and Technology* 2013; 10(4): 343–362.
7. Kumar N., Singh T., Rajoria R.S., and Patnaik A. Optimum design of brake friction material using hybrid entropy-GRA approach. *MATEC Web of Conferences* 2016; 57: 03002.
8. Singh T., Patnaik A., Chauhan R., and Chauhan P. Selection of brake friction materials using hybrid analytical hierarchy process and vise kriterijumska optimizacija kompromisno resenje approach. *Polymer Composites* 2016. doi:10.1002/pc.24113.
9. Singh T., Patnaik A., Chauhan R., Chauhan P., and Kumar N. Physico-mechanical and tribological properties of nanoclay filled friction composite materials using Taguchi design of experiment approach. *Polymer Composites* 2016. doi:10.1002/pc.24101.
10. Singh T., Patnaik A., and Chauhan R. Optimization of tribological properties of cement kiln dust-filled brake pad using grey relation analysis. *Materials and Design* 2016; 89: 1335–1342.
11. Singh T. and Patnaik A. Performance assessment of lapinus-aramid based brake pad hybrid phenolic composites in friction braking. *Archives of Civil and Mechanical Engineering* 2015; 15: 151–161.
12. Singh T., Patnaik A., Chauhan R., and Rishiraj A. Assessment of braking performance of lapinus-wollastonite fibre reinforced friction composite materials. *Journal of King Saud University: Engineering Sciences* 2015. doi:10.1016/j.jksues.2015.06.002.
13. Kumar M. and Bijwe J. Optimized selection of metallic fillers for best combination of performance properties of friction materials: A comprehensive study. *Wear* 2013; 303: 569–583.
14. Bijwe J. and Kumar M. Optimization of steel wool contents in non-asbestos organic (NAO) friction composites for best combination of thermal conductivity and tribo-performance. *Wear* 2007; 263: 1243–1248.
15. Kumar M. and Bijwe J. Role of different metallic fillers in non-asbestos organic (NAO) friction composites for controlling sensitivity of coefficient of friction to load and speed. *Tribology International* 2010; 43(5–6): 965–974.
16. Jang H., Ko K., Kim S.J., Basch R.H., and Fash J.W. The effect of metal fibers on the friction performance of automotive brake friction materials. *Wear* 2004; 256: 406–414.
17. Kumar M. and Bijwe J. Non-asbestos organic (NAO) friction composites: Role of copper; its shape and amount. *Tribology International* 2010; 43: 965–974.

18. Roubicek V., Raclavska H., Juchelkova D., and Filip P. Wear and environmental aspects of composite materials for automotive braking industry. *Wear* 2008; 265(1–2): 167–175.
19. Bijwe J., Kumar M., Gurunath P.V., Desplanques Y., and Degallaix G. Optimization of brass contents for best combination of tribo-performance and thermal conductivity of non-asbestos organic (NAO) friction composites. *Wear* 2008; 265: 699–712.
20. Baklouti M., Cristol A.L., Elleuch R., and Yannick D. Brass in brake linings: Key considerations for its replacement. *Proceedings of the Institution of Mechanical Engineers, Part J: Journal of Engineering Tribology* 2015. doi:10.1177/1350650115588966.
21. Zhu Z., Xu L., Chen G., and Li Y. Optimization on tribological properties of aramid fiber and $CaSO_4$ whisker reinforced non-metallic friction material with analytic hierarchy process and preference ranking organization method for enrichment evaluations. *Material and Design* 2010; 31: 551–555.
22. Milani A.S., Shanian A., and El-Lahham C. Using different ELECTRE methods in strategic planning in the presence of human behavioral resistance. *Journal of Applied Mathematics and Decision Sciences* 2006; 2006: 1–19.
23. Chauhan R., Singh T., Thakur N.S., and Patnaik A. Optimization of parameters in solar thermal collector provided with impinging air jets based upon preference selection index method. *Renewable Energy* 2016; 99: 118–126.
24. Singh T., Patnaik A., and Satapathy B.K. Friction braking performance of nanofilled hybrid fibre reinforced phenolic composites: Influence of nanoclay and carbon nanotubes. *NANO* 2013; 8(3): 1–15.
25. Singh T., Patnaik A., Satapathy B.K., Kumar M., and Tomar B.S. Effect of nanoclay reinforcement on the friction braking performance of hybrid phenolic friction composites. *Journal of Materials Engineering and Performance* 2013; 22(3): 796–805.
26. Singh T. and Patnaik A. Thermo-mechanical and tribological properties of multi-walled carbon nanotube filled friction composite materials. *Polymer Composites* 2015. doi:10.1002/pc.23682.

9 Secondary Manufacturing Techniques for Polymer Matrix Composites

Kishore Debnath, M. Roy Choudhury and T.S. Srivatsan

CONTENTS

9.1 INTRODUCTION

In the time period spanning the past few decades, polymer matrix composites have been chosen for use in a spectrum of applications due in essence to the numerous advantages that they have to offer. The physical properties and mechanical properties of most polymer matrix composites are noticeably superior to the conventional engineering materials. Therefore, these advanced materials have been chosen for use as a viable alternative to the conventional engineering materials. The application spectrum for the family of polymer matrix composites has gradually spread through the domain of engineering practice. This is essentially due to the fact that they are low in weight, economical from a cost perspective and importantly easy to manufacture [1]. However, there are few to several issues that hinder the application of polymer matrix composites for use in the industries spanning automotive, aerospace, civil construction and packaging [2]. The poor interfacial bonding between the matrix and the reinforcing fibre(s), degradation of the matrix that occurs at high temperatures and high moisture absorption characteristics are few of the major disadvantages associated with polymer matrix composites. Further, polymer matrix composites are inhomogeneous, anisotropic and highly abrasive in nature. Also, the thermal conductivity of constituents of the composite is adequately low. Due to these characteristics, there is a tendency for damage to be produced during machining of polymer matrix composites. An extensive amount of analytical and experimental analysis has been undertaken with the primary objective of under-standing the fundamentals of machining behaviour of polymer matrix composites. Unlike monolithic metal, the machining behaviour of polymer matrix composites is dependent on the following: (1) properties of both the reinforcing fibre and the polymer matrix, (2) orientation of the reinforcing fibre and (3) matrix-fibre content in the composite [3]. Drilling, milling, turning, grinding and boring are few of the conventional machining techniques that have been extensively used in the composite industry. However, an investigation of the machining behaviour of polymer matrix composites has revealed several challenges and concomitant research opportunities for both scientists and engineers. Most of the polymer matrix composite parts are often fabricated during primary manufacturing, whereas secondary processing, such as machining and joining, becomes essential during final assembly of the composite parts. A pre-requisite for mechanical joining, such as (1) bolting and (2) riveting, is to generate holes in composite parts. Among the various techniques available, con-ventional drilling is a widely accepted and extensively adopted machining technique for the purpose of producing holes in composite laminates [4]. The adaptability and quality of the mechanically fasted parts necessitate the need for both high preci-sion and accuracy of the hole that is produced during drilling. However, making of holes in a polymer matrix composite using the conventional drilling technique can be safely categorised as a challenging task because damage is produced both in and around the machining zone. The damage is produced primarily due to the genera-tion of torque and thrust force during drilling [5]. The different types of damage that occur during drilling of polymer matrix composites are the following: (1) matrix burning, (2) delamination, (3) chipping, (4) spalling and (5) fibre pull-out, to name a few. A premature failure of the composite structure may result in damage produced during drilling. The rejection of composite parts goes up to 60% as a consequence of the damage produced during drilling in the aircraft industry [6].

9.2   SECONDARY MANUFACTURING

Polymer matrix composites are generally made to near-neat shape during their primary manufacturing, that is, injection moulding, compression moulding and filament winding. However, secondary manufacturing becomes essential for the purpose of making intricate composite structures. A final assembly of various individual parts becomes necessary for the development of an intricate shape composite structure. The assembly of composite parts is often done using mechanical fasteners, such as bolts and rivets. Therefore, making of holes in polymer matrix composites is a vital machining operation that facilitates the use of mechanical fasteners. Among the various techniques available for making holes in a polymer matrix composite, conventional drilling is preferentially chosen and used machining operation. In drilling, a rotating cutting edge advances along its axis of rotation to produce holes in the chosen polymer matrix composites. Section 9.3 discusses the influence of different controlling parameters specific to the drilling operation on process outcomes and quality of the drilled hole with reference to polymer composites.

9.3   CONTROLLING PARAMETERS IN DRILLING OF COMPOSITES

Conventional drilling of polymer matrix composites is often a challenging task when compared to the conventional materials. Therefore, it is significant to recognise the correlation between the drilling parameters in order to analyse their influence on quality of the drilled hole. Through the years, several research investigations have been carried out with the intent to reducing the drilling-induced forces by optimising the (1) drill point geometry, (2) optimising the cutting parameters and (3) selecting the appropriate drill materials. Rakesh and co-workers [7] reported that around 182 articles have been published during the past two decades in the domain of drilling of polymer matrix composites. The researchers have given a major emphasis on the minimisation of damage produced during drilling of composites by optimising the controlling parameters, such as: (a) drill point geometry, (b) cutting speed and feed and (c) tool material.

9.3.1   DRILL POINT GEOMETRY

Drill point geometry is a significant controlling parameter that affects both the thrust force and torque, and subsequently the damage induced during the drilling operation. The traditional twist drill bit is the most common drill bit used for making holes in engineering materials [8]. This type of drill bit has one chisel edge, two primary cutting edges and two marginal cutting edges as shown in Figure 9.1. The other features of drill bit include helix angle, point angle, chisel edge angle, rake angle and clearance angle. A smaller point angle is preferable for making holes in polymer matrix composites. Chen [9] reported that the thrust force increases with an increase in the point angle of the drill bit. Jain and Young [10] observed that the chisel edge of the drill bit contributes roughly about 50 percentage of the total thrust force during the drilling process. In order to reduce the forces generated during drilling, the chisel edge is chosen to be sharp. Consequently, the point angle should be low. The suitable point angle, helix angle and clearance angle of a drill bit for the purpose of making holes in a polymer matrix composite are 60°–135°, 10°–36° and 6°–16°, respectively [11–14].

A conventional twist drill bit is not suitable for producing holes in a polymer matrix composite primarily because drilling with a twist drill bit results in severe damage to the drilled hole. The twist drill bit has a point angle of 118°. The chisel edge of the twist drill bit creates an indention effect during drilling of the composites. Few investigations have been carried out to design and develop different types of drill point geometries with the primary intent of producing damage-free holes in polymer matrix composites. This is shown in Figure 9.2. The straight-flute drill bit [15–18], core drill bit [19–22] and step-core drill bit [23] have been developed, and their performance has been investigated from the context of forces induced and from

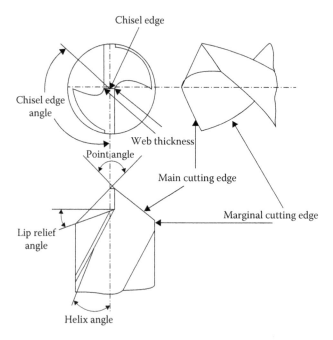

FIGURE 9.1   Nomenclature of the twist drill geometry.

the resultant damage that is produced during drilling of these composites. Piquet and co-workers [15] revealed the damage produced using a twist drill bit to be signifi- cantly high when compared to other developed drill bits during the drilling of carbon fibre-reinforced epoxy composites. Mathew and co-workers [24] reported that the hollow drill bits, such as the trepanning tool, produce minimum force during drilling of the glass fibre-reinforced epoxy composites when compared to the solid twist drill bit. During drilling with the trepanning tool, the cutting action begins from the outer edge of the hollow cylindrical shank, which renders the fibre in tension while con- currently enabling the production of good quality holes. Bajpai and co-workers [8] found that trepanning tool generates a lower thrust force during drilling of sisal fibre- reinforced polypropylene composites and consequently produces less damage than that of the twist drill bit. Miller [25] investigated the drilling behaviour of graphite fibre-reinforced epoxy composite using 17 different types of drill bits. It was found that 8-facet drill bit produces superior quality holes with minimal to no wear of the tool. Singh and Bhatnagar [26] reported that a jo-drill and 8-facet drill bit perform better than a parabolic and 4-facet drill bit during drilling of unidirectional glass fibre-reinforced epoxy composites. However, Lin and Chen [27] pointed out that drilling with multi-facet drill point geometry resulted in rapid wear of the tool when compared to a twist drill bit. The drilling behaviour of sisal fibre-reinforced epoxy and sisal fibre-reinforced polypropylene composites were analysed using three dif- ferent drill bits [28]. The experimental results revealed the overall quality of the hole that is produced by the parabolic drill bit to be superior to the 4-facet and the step drill bit. Velayudham and Krishnamurthy [29] in their independent study found that

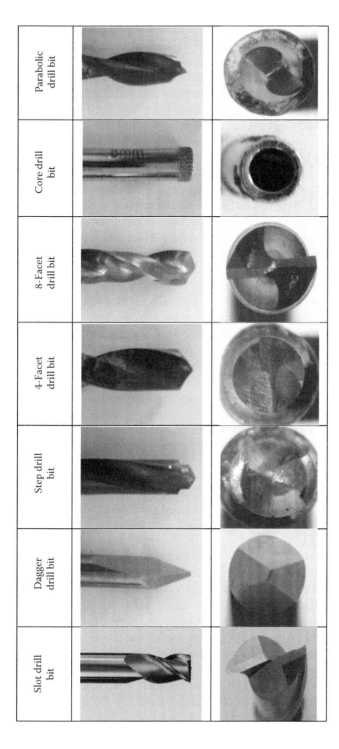

FIGURE 9.2 Specifically designed drill bits for composites.

the tripod drill bit performs relatively better than the thin-tipped and normal-tipped drill bit. Bajpai and co-workers [30] studied the drilling behaviour of green composite laminates by analysing the forces induced and damage using three different drill bits, namely (1) twist, (2) jo and (3) parabolic drill bit. These researchers have reported that the forces generated during drilling with the jo drill bit were higher when compared to the parabolic and twist drill bit. Another study revealed that the jo drill bit produces extensively high thrust force as compared to the 8-facet drill bit during drilling of glass fibre-reinforced vinyl ester composites filled with silicon carbide and graphite powder [31]. The hole-making performance of the nettle fibre-reinforced polypropylene composites using three different drill bits, namely (1) parabolic, (2) 4-facet and (3) step drill bit, has also been investigated [1]. From this experimental investigation, it was found that there occurs a linear increase in the thrust force with an increase in the feed when the drilling is done using a 4-facet and step drill bit. A non-linear relationship was established between the thrust force and feed when drilling is done using the parabolic drill bit. It was also found in the published literature that the parabolic drill bit produces a lower thrust force and torque when compared one-on-one with the other two types of drill bits used. The *straight shank* drill bit resulted in lower cutting stress and power when compared to *brad and spur* drill bit. However, the damage around the surface of the hole was observed to be noticeably more with the *straight shank* drill bit [32,33]. Another independent investigation revealed that the core drill bit generates maximum critical feed among the four different drill bits used, namely (1) core, (2) candlestick, (3) saw and (4) step drill bit [22]. The candlestick drill bit offered minimum critical thrust force when compared to the conventional twist and saw drill bit [34]. Tsao [35] found that the damage produced during drilling using a *step-core-saw* drill bit was less when compared to the *step-core-candlestick* and conventional twist drill bit. Marques and co-workers [36] recorded that the drill bit that generates a minor indentation can produce damage-free holes in a carbon fibre-reinforced epoxy composite. New drill point geometry has been recently developed, which can significantly reduce the forces generated during making of holes in composite laminates [37]. The modified drill bit is composed of two diametrically opposite cutting edges. The damage that is induced during drilling of the polymer composites is significantly reduced when drilling is performed using this new drill bit.

9.3.2 Feed Rate and Cutting Speed

The cutting parameters, namely feed rate and cutting speed, are also significant factors that exert an influence on the forces induced and on the concomitant damage produced during the drilling of polymer matrix composites. In Figures 9.3 and 9.4, the variation of thrust force with feed and torque with feed for three different levels of the spindle speed during drilling of the glass fibre-reinforced epoxy composites using a 4-facet drill bit are shown. From Figures 9.3 and 9.4, it is clear that both thrust force and torque increase with an increase in feed from 0.05 to 0.19 mm/rev. for all values of cutting speed. This can be attributed to the fact that as the feed gradually increases, the thickness or cross-sectional area of the uncut, or undeformed, chip tends to increase. The feed of the drill bit is one vital parameter that helps in determining the overall quality of the hole drilled in the chosen polymer matrix

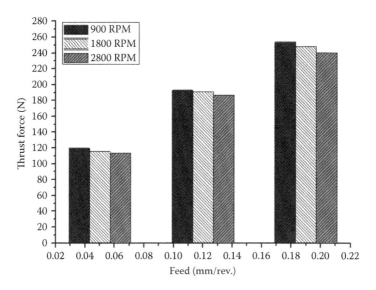

FIGURE 9.3 Variation of thrust force with cutting parameters.

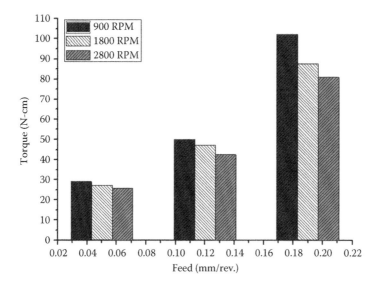

FIGURE 9.4 Variation of torque with cutting parameters.

composite. A damage-free hole in a polymer matrix composite can be obtained by optimal control of the feed rate. Few studies have shown that the forces induced during drilling are more at the higher feed [28,30].

The research investigation showed that the higher spindle speed resulted in relatively lower forces that are induced during the drilling operation [30]. Davim and co-workers [33] observed that the cutting pressure decreases with an increase in spindle speed during drilling of glass fibre-reinforced composites. Consequently, the damage tends to reduce at a higher speed of the drill bit. However, at a high spindle speed the thermal damage experienced by the drilled hole is noticeably more. Sardinas and co-workers [38] and Kilickap [39] observed that the delamination was more at a higher speed of the drill bit. Ramulu and co-workers [40] established the fact that at a high drilling speed the life of the tool tends to decrease. Lin and co-workers [12] noticed that the higher drilling speed produces a higher wear of the tool. At a low spindle speed, ductile mode of chip formation is favoured to occur producing sub-critical cracks and crake-free holes. On the contrary, at a higher spindle speed, the chips tend to form essentially due to brittle fracture of the composite constituents. Hence, it is necessary to set the spindle speed at an optimum value with the primary intent of minimising the damage caused to the hole. According to two independent studies, the suitable cutting speed for polymer matrix composites varies from 78 to 155 mm/m [13,14].

The drilling-induced damage is a function of tool wear rate. At a high spindle speed, the wear of the drill bit is noticeably significant. Wear of the tool is predominant when drilling is done at a high speed. At a high speed, the temperature during machining reaches an elevated temperature, which is conducive for enabling rapid wear of the tool. Hasegawa and co-workers [41] analysed the dependency of tool wear on input parameters. The tool wear was found to be negligible at a low speed. However, a rapid increase in wear of the tool was recorded at a higher spindle speed. Therefore, it can be concluded that a higher spindle speed produces a lower thrust force and the resultant better surface finish while reducing life of the tool.

9.3.3 Tool Materials

The materials such as cemented carbides (coated and uncoated), ceramics, cubic boron nitride (CBN) and diamond (single crystal or polycrystalline) have often been used as drill bit materials for making holes in polymer matrix composites. In Figure 9.5, the different tool materials that have been chosen and used for drilling of fibre-reinforced composites are shown. Cemented carbide and diamond tools are generally used as drill bit materials in commercial practices. Ramulu and co-workers [40] and Wang and co-workers [42] reported that a high-speed steel (HSS) drill wears out at a faster rate during drilling of polymer matrix composites when compared one-on-one with the carbide drills. Kim and Ramulu [43] made a comparison between the performance of steel drill bit and cemented carbide drill bit during high-speed drilling of carbon fibre-reinforced plastic (CFRP)/Ti (fibre metal composite) laminates. These researchers have observed excessive tool wear to occur during drilling with a steel drill bit. Coated cemented carbide drill bit was found to be more suitable than the uncoated cemented carbide drill bit for the purpose of making holes in composite laminates [18]. Davim and Reis [44] reported that a helical flute cemented carbide (K10) drill bit was more suitable when compared to a helical flute HSS drill bit with the intent of producing damage-free holes in polymer matrix composites. Further, the wear resistance capacity of a carbide drill bit was found to be substantially higher than a HSS drill bit. However, a study by Ho-Cheng and Puw [45] revealed that the HSS drill results in negligible tool wear and the resultant production of high-quality holes in a carbon fibre-reinforced composite. Ramulu and co-workers [46] performed drilling experiments on graphite fibre-reinforced

(a)   (b)   (c)

FIGURE 9.5   Drill materials: (a) Cemented carbide twist drill, (b) TiN-coated twist drill and (c) HSS twist drill.

epoxy composites using a polycrystalline diamond (PCD) tool. These researchers concluded that the resistance to tool wear can be significantly improved by using relatively large diamond grains and having a coarse grade PCD.

9.4   OUTPUT RESPONSES IN DRILLING

From the earlier discussion, it is clear that few researchers have independently investigated the influence of different input parameters on both the forces and damage that are induced during drilling of polymer matrix composites. Therefore, a correlation between the controlling parameters and drilling-induced forces and the resultant damage, surface roughness and chip formation characteristics is discussed in the following sections.

9.4.1   Drilling-Induced Thrust Force and Torque

The forces generated during drilling of composites are (1) thrust force and (2) cutting force or torque. Thrust force is an axial force exerted by the drill bit on the work piece and is measured in Newton (N), whereas torque or cutting force is the tangential force generated during drilling and is measured in Newton-centimetre (N-cm). These two drilling-induced forces can be measured using a dynamometer with the aid of force sensors and data acquisition system. Damage caused to the hole is more when the induced thrust force and torque are on the high side. The chisel edge of the drill bit does contribute to a major portion of the thrust force. The chisel edge aids in the removal of material by extrusion as it has a negative rake angle. The primary cutting edges of the drill bit also contribute to the force induced during drilling as bulk of the chip is removed due to an interaction of the primary cutting edges with the composite. The thrust force produced by the cutting edges increases with an increase in the wear of the tool. Further, an abrasive nature of the fibres does facilitate in rapid wear of the cutting edges. The point angle of the drill bit also influences the forces that are induced during drilling. The thrust force increases and the torque decreases as the point angle of the drill bit increases. Therefore, the point angle of the drill bit should be optimum with the primary intent of minimising the forces induced during drilling and the concomitant damage.

The force signals have been correlated with the physical stages of the drilling operation to understand the mechanisms governing drilling. A correlation between the different stages of the drilling operation and the force signals is shown in Figure 9.6. There are five different stages of drilling operation. These are the following:

1. Pre-drilling (I–II)
2. Indentation (II–III)
3. Cutting (III–IV)
4. Reaming (IV–V)
5. Post-drilling (V–VI)

Pre-drilling is the initial stage of the drilling cycle. In this stage, the drill bits start to rotate, but there is no direct contact of the drill bit with the composite laminate. The pre-drilling stops once the chisel edge of the drill bit touches the composite laminate. In the indentation stage, extrusion or indentation of the composite constituents is favoured to occur by the chisel edge. As a consequence, a sudden increase in thrust force is noticed during this stage. The cutting stage begins as the indentation stage culminates and proceeds till the tip of the drill bit exits the bottom most layer of the lamina. In this stage, the bulk material is removed as a consequence of full engagement of the cutting lips with the laminate. The thrust force reaches its maximum value in the cutting stage. Subsequent to the cutting stage, the force signals gradually decrease. This stage is called reaming stage. The reaming stage culminates once the

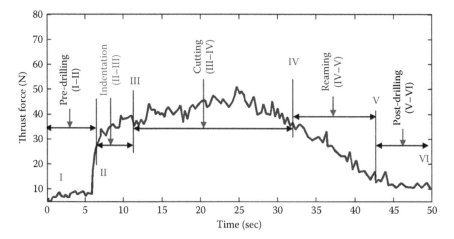

FIGURE 9.6 Correlation of different drilling stages with the force signals.

cutting lip exits the last layer of the laminate. Subsequent to the reaming stage, the force signals get stabilised. This is the last stage of the drilling operation where there is no direct interaction between the drill bit and composite laminate.

Many investigations have been carried out to establish the relationship between the input process parameters and the forces induced during drilling of polymer matrix composites. Abrao and co-workers [47] and Latha and Senthilkumar [48] found that a linear relationship exists between the thrust force and feed rate, and a non-linear relationship exists between the torque and feed rate. Chen [9] investigated the drilling mechanism of unidirectional and multi-directional carbon fibre-reinforced composites using a HSS and carbide drill bit. The relationship between delamination factor and the average thrust force was found to be non-linear during drilling of multi-directional carbon fibre-reinforced composites. This was for the case of the drill bit during drilling of the unidirectional carbon fibre-reinforced composites using a carbide drill bit. This researcher also concluded that composition of the chosen composite material does exert an influence on the drilling mechanism. Hocheng and Tsao [19] observed the area of delamination zone to be determined by the thrust force generated during drilling. During the drilling operation, the value of thrust force below which damage does not occur is called the *critical thrust force*. Rahme and co-workers [49] developed an orthotropic analytical model to analyse the critical thrust force. Di-Ilio and co-workers [50] in their study concluded that both thrust force and torque are associated with specific cutting energy of the drilling operation. The drilling behaviour of fibre reinforced-polymer composites can be improved by establishing the relationships among the various input parameters.

9.4.2 DRILLING-INDUCED DAMAGE

The chosen polymer matrix composite materials are generally anisotropic, inhomogeneous and hydrophilic in nature. Due to these characteristics, the drilling of polymer matrix composites often favours the occurrence of damage in terms of the following: (1) delamination, (2) fibre pull-out, (3) matrix burning, (4) chipping, (5) spalling and (6) microscopic cracking, to name a few. The forces and heat generated due to the intimate contact between the tool and the composite laminate are the primary cause for initiating damage to the hole. Due to different coefficient of thermal expansion of the reinforcing fibre and the polymer matrix, it is quite difficult to attain accurate dimensions of the drilled holes. The holes often tend to shrink subsequent to drilling causing poor tolerance during assembly. Delamination is one of the major forms of damage that occurs during drilling of fibre-reinforced composites. Delamination that is produced during drilling of polymer composites is shown in Figure 9.7. During drilling of unidirectional fibre-reinforced composites, splintering of the fibre is favoured to occur at the entrance or exit of the drill bit. The occurrence of matrix burning, or other thermal-induced damages, can be indicated by visual discolouration of the matrix arising as a consequence of excessive heat generation during drilling. The damage produced during drilling of the polymer matrix composites can be of following types:

1. Delamination
2. Geometric defect
3. Thermal damage

Composite   Delamination

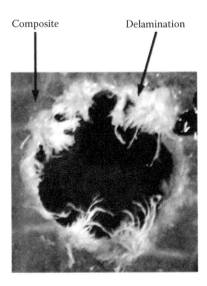

FIGURE 9.7   Induced delamination during drilling of composites.

Several process parameters such as (1) feed rate, (2) spindle speed, (3) drill diameter, (4) drill point design and (5) tool material have a significant influence on drilling-induced damage. The research initiatives that are taken to study the effect of various input parameters on the damage produced during drilling of composites are summarised in Table 9.1.

9.4.3 SURFACE ROUGHNESS

Surface roughness is usually measured using the surface roughness tester and is expressed in terms of microns. Generally, the surface roughness is expressed as arithmetic mean average roughness (R_a). R_a is the arithmetic mean of all vertical deviations from the datum, or the mean line, of the roughness contour and is mathematically expressed as follows:

$$R_a = \int_0^L \left[Y(x) \right] dx \tag{9.1}$$

where:
   R_a is the arithmetic average deviation from the mean line
   Y is the ordinate of the profile curve
   L is the sampling length

The surface roughness of the drilled hole does exert an influence on performance of the final product. Therefore, few studies have been done in an attempt to obtain the minimum surface roughness of the drilled hole by optimising the process

TABLE 9.1

Influence of Various Parameters on the Damage Produced during Drilling of Polymer Matrix Composites

Damage	Material	Tool Material	Significant at	Reference
Delamination at entry	Carbon fibre-reinforced plastic	Cemented tungsten	Higher drill entry feed rate	[32]
Delamination at exit	Carbon fibre-reinforced plastic	Cemented tungsten	Higher cutting speed	[32]
Tool wear, which results in poor surface quality of the drilled hole	Unidirectional carbon fibre-reinforced epoxy composites	Tungsten carbide	Higher cutting speed	[12]
Surface roughness of the hole wall	Glass fibre-reinforced epoxy composites	Cemented carbide	Higher feed rate	[51]
Delamination, inter-laminar cracks, high-density micro-failure areas	E-glass fibre-reinforced polyester composites	HSS	Higher feed rate	[52]
Internal damage of the drilled hole	E-glass fibre-reinforced epoxy composites	–	Angle between the cutting edge and fibre direction is 45° and higher feed rate	[53]
Thermal degradation, reduction of strength, deamination	Carbon fibre-reinforced plastic	Uncoated cemented carbide	Higher speed (higher drilling temperature)	[54]

parameters. Takeyama and Lijima [55] found that surface roughness of the drilled hole wall increases with an increase in cutting speed. They attributed this to the generation of high temperatures during drilling. The temperature generated during drilling results in softening of the polymer matrix, which in turn results in damage of the drilled hole. Orientation of the reinforcing fibre also has a significant influence on roughness value of the machined surface. The maximum surface roughness is obtained at locations of the cutting edge of 135° and 315° and a corresponding fibre orientation of 135° [56]. At this particular orientation of the reinforcing fibres, they are subjected to compression and bending loads. As a consequence of this, fracture of the reinforcing fibre is favoured to occur as a result of both compression and shearing. The surface roughness values obtained during drilling of fibre-reinforced composites are in the range from 4 to 8 μm. The arithmetic average surface roughness value for the carbon fibre-reinforced thermoplastic composites was found to be less than 1 μm [57]. The surface roughness value obtained during machining of carbon fibre-reinforced thermosetting composites (less than 3 μm) is higher than carbon fibre-reinforced thermoplastic polymer composites. Hence, it can be concluded that the surface roughness values obtained during machining of the thermoset-based polymer composite is more than the thermoplastic composite counterpart.

TABLE 9.2
Mode of Chip Formation for Different Composites during Drilling

Composite Material	Tool Martial	Drill Geometry	Chip Formation	Reference
Coir fibre-reinforced polyester composites	HSS (M42)	Twist drill	Continuous chips at low cutting speed and discontinuous chips at high feed rate	[58]
Sisal fibre-reinforced polypropylene composites	Cemented carbide (K10)	Parabolic drill	Continuous chips	[28]
Sisal fibre-reinforced epoxy composites	Cemented carbide (K10)	Parabolic drill	Discontinuous chips	[28]

9.4.4 CHIP FORMATION

The quality of the drilled hole can be predicted by investigating the nature and types of chip formed during drilling of polymer composites. The type of chip formed depends upon both the drilling parameters and the material to be drilled. The chip can be formed as a consequence of two modes: (1) ductile mode and (2) brittle mode. In the ductile mode, the chip generates in the form of a continuous ribbon. In the brittle mode, the chip forms as small fragments. The formation of a continuous chip is favoured to occur during drilling of ductile materials as they do not rapidly fracture. The ductile mode of chip formation is desired to obtain holes without any macroscopic cracks and fine microscopic cracks. The brittle mode of chip formation is not favoured because it is prone to cause severe damage to the drilled hole. At a higher spindle speed, the chip forms in brittle mode, whereas at a lower spindle speed, a continuous chip is favoured to form. The ductile and brittle modes of chip formation also depend on the polymer system chosen. The chip formation modes for different composites are summarised in Table 9.2.

9.5 REMEDIES

The selection of suitable drilling method and optimum drilling condition may result in engineering good quality holes in polymer matrix composites. Enemuoh and co-workers [59] introduced a comprehensive approach to obtain the optimum drilling condition. Damage-free surface can be obtained by monitoring tool wear and by identifying the significant input parameters. Erkki [60] discussed the condition monitoring techniques to reduce tool wear during drilling of composites. It was concluded that Fast Fourier and Wavelet Transform are the two powerful methods that can be used to detect wear of the tool and the resultant damage produced both in and around the drilled hole. An approach that is often used in industries to prevent delamination during drilling is backup support underneath the laminates. The composite laminates often tend to deflect during the drilling operation due to the bending caused by the induced thrust force. This can be prevented by providing a backup support under the laminates. Capello [61] reported that the backup support helps to impede the inflection of composite laminates and consequently reduces the extent of delamination. Tsao and Hocheng in their study [62] noted that the stiffness of the backup support should be sufficiently higher than the stiffness of the laminates. If not, the deflection of the laminate cannot be prevented. Hocheng and co-workers [63] reported that the delamination could be reduced by as much as 60–80 percentage using magnetic colloid backup. Other methods such as (1) high-speed drilling, (2) cryogenic drilling, (3) orbital drilling, (4) drilling with pilot hole and (5) use of cutting fluid have also been tried to promote the drilling performance of polymer matrix composites [3]. Making a pilot hole can also substantially reduce the thrust force and damage that occur during the drilling operation. Pilot holes are made on the work piece prior to the main drilling operation for the purpose of easy guidance of the drill bit. In case of the twist drill bit, the size of the pilot hole is same as that of chisel edge length. The size of the pilot hole is the same as radius of inner drill bit for the case of a core drill bit. This technique allows for the making of holes at a higher feed rate and the resultant higher production rate. Park and co-workers [64] introduced a grinding drill bit with a metal bond PCD. This drill bit tends to offer a higher resistance to wear coupled with minimum delamination. The high-speed drilling was also proved to be a convincing method for the purpose of reducing the damage produced during drilling of polymer matrix composites [30]. However, the operating cost for high-speed drilling is more than that of conventional drilling.

From the earlier discussion, it is quite clear that there is no conventional technique presently available that can completely avoid the drilling-induced delamination in polymer matrix composites. Therefore, limitations posed or put forth by the conventional machining processes can be easily overcome by utilising non-conventional machining techniques, such as (1) electro-discharge machining, (2) electro-chemical machining, (3) water jet machining, (4) abrasive water jet machining, (5) laser beam machining and (6) vibration-assisted drilling.

9.6   SUMMARY

Machining of polymer matrix composites has gained a significant research interest due in essence to their extensive applications. However, the machining behaviour of polymer matrix composites is substantially different from that of conventional materials due to their abrasive, anisotropic and heterogeneous nature. Furthermore, the drilling of polymer matrix composites using conventional methods often results in damage of the hole, wear of the tool and an array of circularity defects of the hole. With the growth of drilling technology, a substantial improvement in both tool materials and tool point geometry has been noticed. Selection of suitable drill point geometry materials for drilling and cutting parameters, that is, feed rate and cutting speed, can help in minimising the damage induced during drilling to an observable or noticeable extent. The development of cost-effective and damage-free drilling method continues to be a challenge to the scientists, engineers and researchers. However, the limitations posed by the conventional techniques used in machining can be eliminated by utilising non-conventional sources of energy or non-conventional machining techniques.

REFERENCES

1. Debnath, K., Singh, I., and Dvivedi, A. 2017. On the analysis of force during secondary processing of natural fiber reinforced composite laminates. *Polymer. Compos.* 38(1):164–174.
2. Bajpai, P.K., Singh, I., and Madaan, J. 2014. Development and characterization of PLA-based green composites: A review. *J. Thermoplast. Compos. Mater.* 27(1):52–81.
3. Komanduri, R. 1997. Machining of fibre-reinforced composites. *Mach. Sci. Technol.* 1(1):113–152.
4. Bhatnagar, N., Singh, I., and Nayak, D. 2004. Damage investigation in drilling of glass fiber reinforced plastic composite laminates. *Mater. Manuf. Process.* 19(6):995–1007.
5. Debnath, K., Singh, I., Dvivedi, A., and Kumar, P. 2013. Recent advances in composite materials for wind turbine blades. In *Advances in Materials Science and Applications*, pp. 25–40. Hong Kong: World Academic Publishing.
6. Stone, R. and Krishnamurthy, K. 1996. A neural network thrust force controller to minimize delamination during drilling of graphite-epoxy laminates. *Int. J. Mach. Tools Manuf.* 36:985–1003.
7. Rakesh, P.K., Singh, I., and Kumar, D. 2012. Drilling of composite laminates with solid and hollow drill point geometries. *J. Compos. Mater.* 46:3173–3180.
8. Bajpai, P.K. and Singh, I. 2013. Drilling behaviour of sisal fibre reinforced polypropylene composite laminates. *J. Reinf. Plast. Compos.* 32(20):1569–1576.
9. Chen, W.C. 1997. Some experimental investigation in the drilling of carbon fibre reinforced plastic (CFRP) composite laminate. *Int. J. Mach. Tools Manuf.* 37:1097–1108.
10. Jain, S. and Yang, D.C.H. 1993. Effects of feed rate and chisel edge on delamination in composite drilling. *Trans. ASME J. Eng. Ind.* 115(4):398–405.
11. Liu, D., Tang, Y., and Cong, W.L. 2012. A review of mechanical drilling for composite laminates. *Compos. Struct.* 94(4):1265–1279.
12. Lin, S.C. and Chen, I.K. 1996. Drilling of carbon fibre—Reinforced composite material at high speed. *Wear* 194(1–2):156–162.
13. Kim, D., Doan, X., and Ramulu, M. 2005. Drilling performance and machinability of PIXA-M and PEEK thermoplastic composites. *J. Thermoplast. Compos. Mater.* 18(3):195–217.
14. Zitoune, R., Krishnaraj, V., Almabouacif, S., Collombet, F., Sima, M., and Jolin, A. 2012. Influence of machining parameters and new nano-coated tool on drilling performance of CFRP/aluminium sandwich. *Compos. B.* 43(3):1480–1488.
15. Piquet, R., Ferret, B., Lachaud, F., and Swider, P. 2000. Experimental analysis of drilling damage in thin carbon/epoxy plate using special drills. *Compos. A.* 31:1107–1115.
16. Fernandes, M. and Cook, C. 2006. Drilling carbon composites using a one shot drill bit. Part I. Five stage representation of drilling and factor affecting maximum force and torque. *Int. J. Mach. Tools Manuf.* 46:70–75.
17. Fernandes, M. and Cook, C. 2006. Drilling carbon composites using a one shot drill bit. Part II. Empirical modelling of maximum thrust force. *Int. J. Mach. Tools Manuf.* 46:70–79.
18. Murphy, C., Byrne, G., and Glichrist, M.D. 2002. The performance of coated tungsten carbide drills when machining carbon fibre-reinforced epoxy composite materials. *Proc. Inst. Mech. Eng., B J. Eng. Manuf.* 216:143–152.
19. Hocheng, H. and Tsao, C.C. 2003. Comprehensive analysis of delamination in drilling of composite materials with various drill bits. *J. Mater. Process. Technol.* 140(1):335–339.
20. Tsao, C.C. and Hocheng, H. 2007. Parametric study on thrust force of core drill. *J. Mater. Process. Technol.* 192–193:37–40.
21. Jain, S. and Yang, C.C. 1994. Delamination-free drilling of composite laminates. *ASME J. Eng. Ind.*, 116:475–481.

22. Hocheng, H. and Tsao, C.C. 2006. Effects of special drill bits on drilling-induced delamination of composite materials. *Int. J. Mach. Tools Manuf.* 46(12):1403–1416.
23. Tsao, C.C. 2008. Experimental study of drilling composite with step-core drill. *Mater. Des.* 29:1740–1744.
24. Mathew, J., Ramakrishnan, N., and Naik, N.K. 1999. Investigations into the effect of geometry of a trepanning tool on thrust and torque during drilling of GFRP composites. *J. Mater. Process. Technol.* 91(1):1–11.
25. Miller, J.A. 1987. Drilling graphite/epoxy at Lockheed. *Am. Mach. Auto. Manuf.* 131:70–71.
26. Singh, I. and Bhatnagar, N. 2006. Drilling of uni-directional glass fiber reinforced plastic (UD-GFRP) composite laminates. *Int. J. Adv. Manuf. Technol.* 27(9–10):870–876.
27. Lin, S.C. and Chen, J.M. 1999. Drilling unidirectional glass fibre-reinforced composite materials at high speed. *J. Compos. Mater.* 33(9):827–851.
28. Debnath, K., Singh, I., and Dvivedi, A. 2014. Drilling characteristics of sisal fibre-reinforced epoxy and polypropylene composites. *Mater. Manuf. Process.* 29:1401–1409.
29. Velayudham, A. and Krishnamurthy, R. 2007. Effect of point geometry and their influence on thrust and delamination in drilling of polymeric composites. *J. Mater. Process. Technol.* 185(1):204–209.
30. Bajpai, P., Debnath, K., and Singh, I. 2017. Hole making in natural fibre-reinforced polylactic acid laminates: An experimental investigation. *J. Thermoplast. Compos. Mater.* 30:30–46.
31. Kumar, S., Chauhan, S.R., Rakesh, P.K., Singh, I., and Devim, J.P. 2008. Drilling of glass fibre/vinyl ester composites with fillers. *Mater. Manuf. Process.* 27:314–319.
32. Davim, J.P. and Reis, P. 2003. Drilling carbon fiber reinforced plastics manufactured by autoclave: experimental and statistical study. *Mater. Des.* 24(5):315–324.
33. Davim, J.P., Reis, P., and Antonio, C.C. 2004. Experimental study of drilling glass fibre reinforced plastics (GFRP) manufactured by hand lay-up. *Compos. Sci. Technol.* 64(2):289–297.
34. Tsao, C.C. and Hocheng, H. 2005. Computerized tomography and C-Scan for measuring delamination in the drilling of composite materials using various drills. *Int. J. Mach. Tools Manuf.* 45(11):1282–1287.
35. Tsao, C.C. 2008. Investigation into the effects of drilling parameters on delamination by various step-core drills. *J. Mater. Process. Technol.* 206(1):405–411.
36. Marques, A.T., Durão, L.M.P., Magalhães, A.G., Silva, J.F., and Tavares, J.M.R. 2009. Delamination analysis of carbon fibre reinforced laminates: evaluation of a special step drill. *Compos. Sci. Technol.* 69(14):2376–2382.
37. Debnath, K., Sisodia, M., Kumar, A., and Singh, I. 2016. Damage-free hole making in fibre-reinforced composites: An innovative tool design approach. *Mater. Manuf. Process.* 31:1400–1408.
38. Sardinas, R.Q., Reis, P., and Davim, J.P. 2005. Multi-objective optimization of cutting parameters for drilling laminate composite material by using genetic algorithms, *Compos. Sci. Technol.* 65:455–466.
39. Kilickap, E. 2010. Optimization of cutting parameters on delamination based on Taguchi method during drilling of GERP composite. *Expert Syst. Appl.* 37:6116–6122.
40. Ramulu, M., Branson, T., and Kim, D. 2001. A study on the drilling of composite and titanium stacks. *Compos. Struct.* 54(1):67–77.
41. Hasegawa. Y., Hanasaki, S., and Satanaka, S. 1984. Characteristics of tool wear in cutting of GFRP. *Proceedings of the 5th International Conference on Production Engineering*, Tokyo, Japan, pp. 185–190.
42. Wang, X., Wang, L. J., and Tao, J. P. 2004. Investigation on thrust in vibration drilling of fibre reinforced plastics. *J. Mater. Process. Technol.* 148(2):239–244.

43. Kim, D. and Ramulu, M. 2004. Drilling process optimization for graphite/ bismaleimide-titanium stacks. *Compos. Struct*, 63:101–114.

44. Davim, J.P. and Reis, P. 2003. Study of delamination in drilling carbon fibre reinforced plastics (CFRP) using design experiments. *Compos. Struct*. 59(4):481–487.

45. Ho-Cheng, H. and Puw, H.Y. 1993. Machinability of fibre-reinforced thermoplastics in drilling. *J. Eng. Mater. Technol*. 115:146–149.

46. Ramulu, M., Faridnia, M., Garbini, J.L., and Jorgensen, J.E. 1991. Machining of graphite/epoxy composite materials with polycrystalline diamond (PCD) tools. *J. Eng. Mater. Technol*. 113:430–436.

47. Abrao, A.M., Rubio, J.C., Faria, P.E., and Davim, J.P. 2008. The effect of cutting tool geometry on thrust force and delamination when drilling glass fibre reinforced plastic composite. *Mater. Des*. 29(2):508–513.

48. Latha, B. and Senthilkumar, V.S. 2009. Analysis of thrust force in drilling glass fibre-reinforced plastic composites using fuzzy logic. *Mater. Manuf. Process*. 24(4):509–516.

49. Rahme, P., Landon, Y., Lachaud, F., Piquet, R., and Lagarrigue, P. 2011. Analytical models of composite material drilling. *Int. J. Adv. Manuf. Technol*. 52(5–8):609–617.

50. A. Ilio, Di, Tagliaferri, V., and Veniali, F. 1993. Specific cutting energy in drilling of composites. *Proceedings of the Symposium on Machining of Advanced Composites*, edited by M. Ramulu and R. Komanduri, ASME, MD-45IPED-66, pp. 53–64.

51. Ogawa, K., Aoyama, E., Inoue, H., Hirogaki, T., Nobe, H., Kitahara, Y., Katayama, T., and Gunjima, M. 1997. Investigation on cutting mechanism in small diameter drilling for GFRP (thrust force and surface roughness at drilled hole wall). *Compos. Struct*. 38(1–4):343–350.

52. Caprino, G. and Tagliaferri, V. 1995. Damage development in drilling glass fibre reinforced plastics. *Int. J. Mach. Tools Manuf*. 35(6):817–829.

53. Aoyama, E., Nobe, H., and Hirogaki, T. 2001. Drilled hole damage of small diameter drilling in printed wiring board. *J. Mater. Process. Technol*. 118(1–3):436–441.

54. Yashiro, T., Ogawa, T., and Sasahara, H. 2013. Temperature measurement of cutting tool and machined surface layer in milling of CFRP. *Int. J. Mach. Tools Manuf*. 70:63–69.

55. Takeyama, H. and Lijima, N. 1988. Machinability of glass fibre reinforced plastics and application of ultrasonic machining. *Ann CIRP*. 37(1):93–96.

56. Konig, W., Wolf, C., Grab, P., and Willerscheid, H. 1985. Machining of fibre reinforced plastics. *Ann CIRP*. 34:537–547.

57. Hocheng, H. and Puw, H.Y. 1992. On drilling characteristics of fibre-reinforced thermoset and thermoplastics. *Int. J. Mach. Tools Manuf*. 32:583–592.

58. Jayabal, S. and Natarajan, U. 2011. Drilling analysis of coir-fibre reinforced polyester composites. *Bull. Mater. Sci*. 34(7):1563–1567.

59. Enemuoh, E.U., El-Gizawy, A.S., and Okafor, A.C. 2001. An approach for development of damage-free drilling of carbon fibre reinforced thermosets. *Int. J. Mach. Tools Manuf*. 41(12):1795–1814.

60. Erkki, J. 2002. A summary of methods applied to tool condition monitoring in drilling. *Int. J. Mach. Tools Manuf*. 42:997–1010.

61. Capello, E. 2004. Workpiece damping and its effect on delamination damage in drilling thin composite laminates. *J. Mater. Process. Technol*. 148(2):186–195.

62. Tsao, C.C. and Hocheng, H. 2005. Effects of exit back-up on delamination in drilling composite materials using a saw drill and a core drill. *Int. J. Mach. Tools Manuf*. 45(11):1261–1270.

63. Hocheng, H., Tsao, C.C., Liu, C.S., and Chen, H.A. 2014. Reducing drilling-induced delamination in composite tube by magnetic colloid backup. *Ann CIRP*. 63(1):85–88.

64. Park, K.Y., Choi, J.H., and Lee, D.G. 1995. Delamination – Free and high efficiency drilling of carbon fibre reinforced plastic. *J. Compos. Mater*. 29:1988–2002.

10 Drilling Behaviour of Natural Fibre-Reinforced Polymer Matrix Composites

Inderdeep Singh, U.K. Komal and M.K. Lila

CONTENTS

10.1 INTRODUCTION

Composites have been proved to be an effective tailor-made material for present day industry due to unique combination of properties, which are not available in traditional materials. The superior advantages (such as light weight, high strength-to-weight ratio, chemical resistance and design flexibility) of composite material have led them to replace the conventional materials in many applications including automotive and aerospace. Fibre-reinforced polymers (FRPs) are the most common and widely used composite materials, which offers numerous advantages over conventional materials, that is, high strength-to-weight ratio, corrosion resistance, lower maintenance costs, ease of processing and transportation and the ability to produce near net shape products.

Currently, synthetic fibres are extensively being used as reinforcement in the fibre-reinforced polymer composites in structural applications. Glass, aramid and carbon fibres are commonly used synthetic fibres. Synthetic FRPs have shown the structural durability in all types of interior and exterior applications (Holbery and Houston 2006). Good mechanical properties have facilitated the entry of these materials in many applications; areas vary from the highly sophisticated aerospace industry to the more common sporting goods. Although these composites possess excellent mechanical properties, their major drawbacks are poor machinability and recycling characteristics along with the non-biodegradable nature, which is a major threat to ecological system. Requirement of lighter products and strict environmental policies and regulations in many engineering applications have led to the development of natural fibre-reinforced polymer composites (NFRPCs).

In the past decade, the feasibility of using natural fibres as reinforcement for thermoplastic and thermoset matrices has been explored and successfully adopted by various automobile manufacturers and suppliers for door panels, seat backs, headliners, package trays, dashboards and so on. Natural fibres have attracted the attention of researchers, engineers and scientists as an alternative reinforcement material for NFRPCs due to their excellent properties, that is low density, high modulus, non-abrasive nature, ease of fibre surface modification, abundant availability and most important, environment friendliness. Similar to synthetic fibre-reinforced composites, NFRPCs can also be engineered easily to meet specific requirement of products for different applications (Najaf et al. 2012). These have encouraged the automotive and aerospace industry to use natural fibres as alternative reinforcing materials in their polymeric parts. NFRPCs have shown comparative or even better characteristics than synthetic fibre-reinforced polymer composites. NFRPC are either partially or fully biodegradable, depending upon the polymeric-matrix materials. When natural fibres are used to reinforce non-biodegradable petroleum-based polymers, such as polypropylene, polyethylene and so on, then they are called as partially biodegradable composites but are called as fully biodegradable, when natural fibres are used to reinforce biodegradable polymers, that is, polylactic acid (PLA), cellulosic plastic and so on. Figure 10.1 shows the constituents of NFRPC.

The properties and performance of NFRPC products mainly depend on their processing methods and input parameters, which involve two steps: primary and secondary processing. Primary processing is often termed for fabricating the composites,

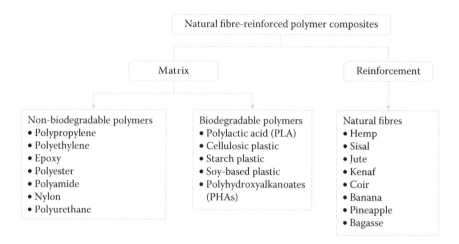

FIGURE 10.1   Constituents of NFRPCs.

that is, hand lay-up, compression moulding, injection moulding and so on, which are already well-established methods. Secondary processing is the essential and the second step to fabricate the final composite products. In order to fabricate high-performance composites, it is necessary to focus on various aspects of the secondary processing techniques. This chapter provides a brief discussion on various secondary processing techniques, challenges and opportunities and the process parameters, which influence the properties and performance of the polymer composites.

10.2   SECONDARY PROCESSING OF NATURAL FIBRE-REINFORCED POLYMER COMPOSITES

10.2.1   THE NEED FOR SECONDARY PROCESSING

Although most of the polymer composite products are processed to a near-net shape, several parts and components have to be joined together to assemble the final product in order to fabricate intricate products for certain applications. There are a wide variety of joining techniques (adhesive bonding, microwave joining, mechanical fastening etc.) available to assemble the individual composite parts. Adhesive bonding and novel microwave joining of polymer composites are not always a feasible option to get the structural integrity of the assembly. Joining by means of adhesive bonding is not suitable in the case where temporary joining is required and novel microwave joining technique is limited to the thermoplastic-based composite parts. These problems can be overcome by using mechanical fastening for joining and assembling the individual parts. For mechanical fastening, certain machining operations such as drilling, trimming and finishing are required. Among these machining processes, drilling is the most commonly and frequently used method for making holes in order to assemble the composite parts. It is well understood that the efficiency of the mechanical fastening is highly dependent on the quality and accuracy of the drilled hole (Debnath et al. 2015a).

10.2.2 DRILLING OF NATURAL FIBRE-REINFORCED POLYMER COMPOSITES

NFRPCs have been proved and are being used as excellent materials for non-structural applications. The drilling of these materials is entirely different as compared to drilling of metals because of their anisotropic nature; due to that, the tool has to cut distinct phases simultaneously. The effect of drilling depends on the various parameters such as properties of polymer matrix and natural fibre, fibre volume fraction and fibre orientation. Drilling of polymer matrix composites causes various kinds of damage within the composite and at its surface (Rakesh et al. 2012). Drilling-induced damage (delamination, deformation, cracking, fibre pull-out and hole ovality) has a severe impact on the structural integrity of composite parts. Drilling process parameters such as cutting speed, feed rate, tool material and tool geometry are some of the main factors, which governs the quality of drilled holes. Optimisation of these parameters, numerical modelling and unconventional methods of hole making are some of the important research area, which have been undertaken to optimise the drilling operation in order to obtain the damage-free holes.

Setup for drilling of composites mainly depends on the output parameters that are required to analyse the effect of drilling on the quality and performance of composites. Thrust force and torque are the two important output factors. A typical drilling experimental setup consists of drilling machine, work-holding fixture, piezoelectric dynamometer, charge amplifier, an analogue to digital convertor and a computer for data acquisition. Dynamometer is used to record the signals of thrust force and torque; amplifier can be used to improve the signal received from the dynamometer. Dedicated data acquisition system and software are used to record and analyse the thrust force and torque signals (Singh et al. 2008). Figure 10.2 shows the schematic of the typical drilling setup.

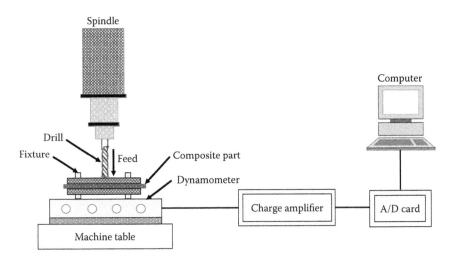

FIGURE 10.2   Schematic of the typical drilling setup.

10.3   ISSUES AND CHALLENGES IN DRILLING OF POLYMER COMPOSITES

A certain degree of intricacy in the composite product design necessitates the joining of individually processed parts into a complex composite product (Bajpai et al. 2012). In order to produce an intricate composite product assembly, several composite parts and components have to be joined together through various techniques, that is, riveting, bolting and so on; therefore, drilling is the essential operation required to fabricate the hole in the composite part for assembly operation. Drilling of polymer composites by conventional methods is a highly challenging task as the damage produced due to unavoidable conditions results in rejection of the composite parts (Debnath et al. 2015b). This has been estimated that most of the fabricated composite parts are rejected in the manufacturing industry due to the drilling-induced damage around the drilled hole. Drilling-induced damage generally occurs due to direct interaction between the drilling tool and composite laminate. This damage is mainly affected by the heat generated during drilling operation, thrust force and sometimes due to poor adhesion between fibre and matrix. Also, abrasive nature of some fibres leads to increased tool wear, which again leads to higher cost of processing the part. Drilling-induced damage is a major concern in secondary operation of composites.

10.3.1   DRILLING-INDUCED DAMAGE

The evaluation of drilling-induced damage is an important area in the machining of polymer composites. The research interest in this area has gained widespread attention in the past few decades. The major emphasis is on the minimisation of drilling-induced damage. Drilling-induced damage in composite laminates can be in the form of delamination, fibre pull-out, hole ovality, matrix burning and cracks along the drilled hole. This damage is responsible for the failure of the composite parts during service life. The nature of the drilling-induced damage depends upon the various parameters such as constituents of the composites, polymer matrices (thermoplastic or thermoset, ductile or brittle), nature of the reinforcement, tool material and geometry and so on. Preventive and quality control measures are essential to detect the damage occurring during drilling in order to ensure the high quality and performance of the finished composite products. The studies on drilling-induced damage have been qualitative as well as quantitative in nature. The damage can be quantified with the help of digital image processing as well as non-destructive testing techniques. The Section 10.3.2 deals with the details of various types of drilling-induced damage.

10.3.2   DELAMINATION

Delamination in composite laminates means the separation of layers or laminae. Delamination can be critical as it may severely affect the integrity of matrix and reinforcement as well as long-term reliability of the components. The major reason is that delamination creates potential points of origin for failure when the product is under loaded conditions. Two different types of delamination can be observed while drilling fibre-reinforced composite laminates, namely peel-up and push-down delamination.

Peel-up delamination takes place during the initiation of the drilling process as the cutting tip comes in contact with the composite. This leads to separation and bending of the surface layers, leading to fracture. A peeling force through the slope of the flutes separates the layers of reinforcement in composite laminate and results in delamination around the entry side of the drilled hole.

When the drill approaches towards the exit point, push-down delamination occurs due to the thrust force, which is generated due to the drill feed (Nassar et al. 2016). When the generated thrust force exceeds the inter-ply bond strength between the matrix and reinforcement; this results in stretching of plies under the drill tool and causes delamination between reinforcement layers, which is termed as push-down delamination. The drill point exerts compressive force on the uncut plies, causing them to bend elastically. As the drill approaches the exit, the number of uncut plies supporting it reduces and the resistance to bending decreases. At a critical thickness, the bending stress becomes greater than the inter-laminar strength between the plies, and an inter-laminar crack is initiated around the hole. Further pushing down by the drill point causes the crack to propagate, and the flexural rigidity of the supporting plies becomes weaker. This leads to fracture of material under the drill point. The damage at exit plies is shown as spalling that extends beyond the hole diameter. The mechanism of both types of delamination (peel-up and push-down) has been shown in Figure 10.3. Both types of delamination result in deviation in the required hole diameter and strength of the material around the hole. In conclusion, there are two types of delamination. First, entry of the drill into the laminate is known as the peel-up type, and delamination during exit is known as push-down delamination (Singh and Bajpai 2013).

Delamination is the most common type of damage that occurs during the drilling of composite laminates. Drilling-induced delamination can be quantified by using various digital processing software such as Image J1.42. A factor known as delamination factor can be used to quantify the drilling-induced damage.

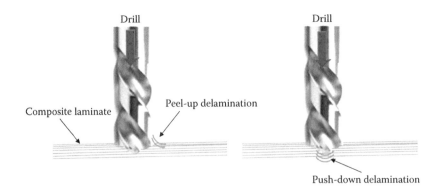

FIGURE 10.3 Peel-up and push-down type of delamination.

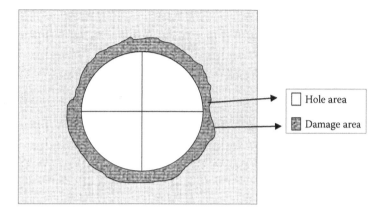

FIGURE 10.4   Schematic representation of hole area and damage area.

$$\text{Delamination factor, } (F_d) = \frac{A_{max}}{A_{hole}} \qquad (10.1)$$

where:

A_{max} is the maximum area (hole area + damage area)
A_{hole} is the hole area = $(\pi/4)D^2$
D is the diameter of hole

A schematic diagram showing A_{max} and A_{hole} is shown in Figure 10.4.

10.3.3   FIBRE PULL-OUT

Fibre pull-out is also a major reason for deterioration of the hole quality and for the drastic reduction in the strength around the hole. This generally happens due to poor interfacial bonding between matrix and reinforcement, which leads to the extraction of fibre during the operations. When thick fibre strands are used in mat form to reinforce the matrix, high viscosity of the matrix material causes poor wetting of the fibres, which leads to lower interfacial bonding between the matrix and reinforcement. Therefore, while drilling, when these fibres comes in contact with the cutting edge and if the torque is higher, debonding occurs and fibres get pulled out partially or completely from the laminates and weaken the area around the hole. Fibre pull-out also results in poor hole quality and strength of the composite laminate around the hole. Figure 10.5 shows the fibre pull-out during drilling operation.

10.3.4   HOLE OVALITY

Hole ovality deals with the shape of the hole with reference to the drill diameter, which generally occurs when the orientation of fibre reinforcement is unidirectional in the composite laminates. In unidirectional fibre-reinforced composite laminates, during drilling operation, the cutting angle varies with the rotation of the drill tool.

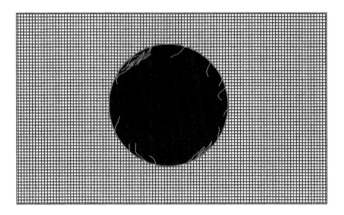

FIGURE 10.5 Fibre pull-out.

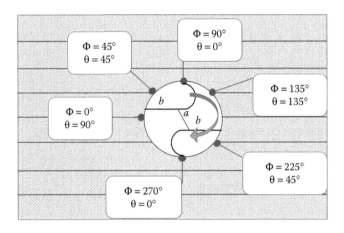

FIGURE 10.6 Variation of cutting angle with respect to the rotation of drill point.

The mechanism is shown in Figure 10.6. It can be observed that when the cutting tool is performing the cutting action, it encounters the fibres at different angles. Therefore, especially in case of unidirectional fibre-reinforced composites, the hole produced is not purely circular but is mostly oval shaped.

10.3.5 MATRIX BURNING

It has been investigated and concluded that the torque and thrust force result in heat generation along the periphery of the hole. More forces result in more heat generation and when this generated heat attains the critical value, the burning of matrix takes place.

10.4 OPPORTUNITIES

The drilling-induced damage during drilling operation has become a prime area of research, and efforts have been made worldwide for minimizing the damage. Apart from changes in various process parameters, such as cutting speed, feed rate, tool materials and geometry, the researchers have attempted several process modifications in the machining operation in order to decrease the drilling-induced damage. The various process modifications attempted by the researchers are as follows:

- Helical feed method
- Backup plate method
- Ultrasonic-assisted drilling

10.4.1 HELICAL FEED METHOD

A helical feed drilling method has been attempted (Figure 10.7) to reduce the drilling-induced damage over conventional drilling. It is also known as the Kungl Tekniska Hogskolan (KTH) method proposed by Zackrisson et al. (1994). In this method, the drill moves down helically with respect to the drilling axis, whereas the drill in the conventional drilling moves straight down along the axis. In the new method, the removal of chips and the supply of the cutting fluid would be better because there is enough gap between the composite part and the tool. A ball-nose type core drill is used in this method.

Advantages of helical feed method:

- Better chip removal
- Decrease in thrust force
- Decrease in drilling-induced damage
- Better flow of coolant
- Less fibre matrix burning due to lower temperature during the operation

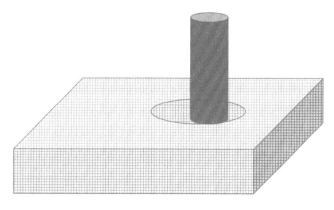

FIGURE 10.7   Helical feed method.

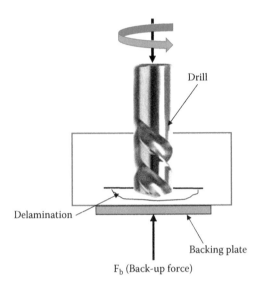

FIGURE 10.8   Use of backing plate.

10.4.2   Backup Plate Method

An approach that is often used in many industries to prevent push-down delamination during drilling is using backup support underneath the composite part being drilled. The backup plate restricts the downward bending deflection of the composite part caused by the thrust force during drilling. While selecting the backup material, it is important to note that the stiffness of the backup plate should be higher than the stiffness of the composite parts; otherwise, the bending deflection of the composite parts caused by thrust force cannot be restricted by the backup support. It should be clear that the backup plate does not provide strength, during operation; it only impedes or arrests the inflection of composite laminates, which causes reduced extent of delamination. The use of the backup plate could marginally reduce the push-down delamination. Figure 10.8 shows the use of backing plate during the operation.

Advantages of backup plate method:

- Reduction in push-down delamination
- Improved quality of hole
- Less burr formation at the bottom (especially in case of thermoplastic-based composites)

10.4.3   Ultrasonic-Assisted Drilling

In order to minimise the drilling-induced damage during the conventional drilling technique, certain innovative drilling approaches have been attempted by the researchers to get good quality holes in composite parts with greater efficiency. Ultrasonic-assisted drilling is an example of such innovative drilling approach.

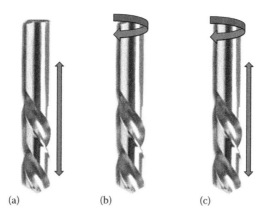

FIGURE 10.9   Ultrasonic-assisted drilling: (a) longitudinal vibration, (b) torsional vibration and (c) combined, longitudinal and torsional vibration.

Ultrasonic vibration can be applied to a drill in several ways (Figure 10.9). Purely longitudinal vibration along the axis of the drill, purely torsional vibration or a combined longitudinal/torsional motion can be applied. Another method is to apply ultrasonic vibration to the work piece during drilling operation (Thomas and Babistky 2007).

Advantages of ultrasonic-assisted drilling:

- Less burr formation
- Improved chip removal
- Reduced-drilling thrust force and torque
- Better quality hole
- Durability of drill bit
- Increased material removal rate

10.5   FACTORS AFFECTING THE DRILLING OPERATION

There are various factors, which affects the quality of hole during drilling operation (Figure 10.10), including tool material, tool geometry, matrix material, reinforcing material as well as input parameters, that is, drill speed and feed rate. All these parameters have direct or indirect influence on variation in output parameters, that is, thrust force and torque, which affects the quality of hole (Rajakumar et al. 2012). Some of these are discussed in Sections 10.5.1 through 10.5.4.

10.5.1   EFFECT OF DRILLING TOOL

Generally, high-speed steel (HSS) drill bits can be used for drilling of natural fibre-reinforced composites, but when considering tool life and quality of holes, other materials, that is, solid carbide and tungsten carbide can also be used.

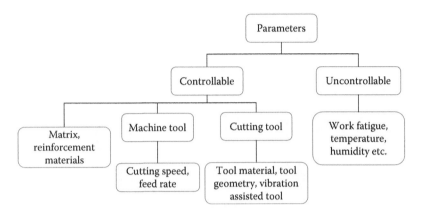

FIGURE 10.10   Factors affecting the drilling of polymer composites.

Generally, natural fibres are non-abrasive in nature, so they do not impart an adverse effect on the tool wear. Drill bit geometry has a relevant effect on thrust force as well as on torque, thus resulting in variation in drilling-induced damages as mentioned earlier. Different types of conventional drill bits, that is, twist drill, parabolic drill, multi-facet drill, jo drill, shaped drill, trepanning tool and so on have been purposed by various researchers for drilling operation based on output in the form of thrust force and torque for different materials under different processing conditions.

Certain modified tool geometries (parabolic, U shaped etc.) have also been proposed by various researchers to improve the quality of holes. Along with these, drill bit diameter also has the major effect on the resulting forces. Therefore, all these have to be optimised in order to reduce the damage associated with drilling operation.

When considering tool geometry, if drilling is performed with twist drill, the drilling starts from the centre of the drill (indentation of the chisel edge on work material) and then there is complete engagement between the drill and work material, whereas in case of trepanning tool, the mechanism is substantially different as cutting action starts from the periphery of trepanning tool, by the cutting edges on the periphery of hollow cylindrical shank (Bajpai and Singh 2013). Debnath et al. (2015a) investigated the influence of three types of drill geometry (step drill, 4-facet drill and parabolic drill) with different feed rates and spindle speeds on nettle fibre-reinforced polypropylene laminates. It was reported that thrust force increases with an increase in the feed rate in case of 4-facet and step drill but shows non-linear relation for parabolic drill. It was also mentioned that the maximum thrust force and torque were generated in the case of step drill and 4-facet drill.

Optimisation studies show that a combination of low feed rate and high spindle speed is required in order to minimise the thrust force and torque; in terms of geometry, parabolic drill should be preferred (Rajakumar et al. 2012). Similar result was also observed in the case of sisal fibre-reinforced epoxy and polypropylene composites. In other investigations by Bajpai et al. (2015), parabolic drill was reported as the best among the twist drill and jo drill for sisal and nettle fibre-reinforced PLA and polypropylene composites. It was also mentioned that the tool geometry has a major influence on cutting forces during drilling operation when compared with other parameters.

10.5.2   Effect of Matrix Material

The mechanical and thermal properties of the matrix material also influence the chip formation process as these properties are directly related to the deformation behaviour. Therefore, drilling of thermosets and thermoplastics results in two different types of chips. Thermosets are stiffer and brittle in nature, which produce chips in powder form due to brittle fracture at tool chip interface during drilling operation. At higher speed and low feed rate, thermoset matrix fractures early because of high strain rate, resulting in smaller chip segments, whereas in thermoplastic matrix, continuous chips (ribbon like) are formed due to their viscoelastic nature. This phenomenon in thermoplastics may be attributed to the elongation under loading condition and slipping among the hydrocarbon chains, due to rise in temperature during drilling operation (Ahmad 2009). The thermal conductivity of thermoplastics is also low, so that local heating and cooling occur at hole boundaries resulting into good quality holes. But the chip morphology may be affected by the reinforcement and cutting conditions.

10.5.3   Effect of Reinforcement

Reinforcing materials also differ in their mechanical, physical and thermal properties, resulting in exhibiting different types of failure modes and surface morphology during drilling operation. Synthetic fibres (carbon, glass) exhibit brittle fracture when compared to natural fibres, which shows ductile failure during drilling operation.

Fibre orientation has more influence on cutting and thrust forces, whereas operating conditions and tool geometry have comparatively less influence. This phenomenon can also be explained with the help of Figure 10.6. An increase in the cutting force is exhibited when cutting angle is 90°, that is, when the cutting edge is perpendicular to the fibre orientation. When the cutting angle is 0°, that is, when the cutting edge is along the fibre orientation, the cutting force decreases.

10.5.4   Effect of Process Parameter

As it is mentioned earlier that the thrust force and torque have direct influence on the quality of hole, these forces mainly depend on the input parameters, that is, cutting speed and feed. The forces generated during drilling are mostly accountable for peel-up and push-down type of delamination in composite laminates. Therefore, the analysis of thrust force and torque is an important consideration because the magnitude and nature of the drilling forces define the quality of the drilled hole (Debnath et al. 2014). Therefore, these input parameters should be optimised before starting the drilling operation in order to achieve good quality hole. Drilling force increases with the feed rate, whereas the speed has a non-linear relation with the cutting forces.

The optimum conditions for drilling are less force generation, less power consumption, minimal tool wear with high material removal rate and good hole quality (without drilling-induced damage). In the studies conducted by the researchers, it has been reported that, optimisation of process parameters by different optimisation techniques can be used to minimise the generation of thrust force and torque.

10.6 OPTIMISATION OF DRILLING PARAMETERS

The response-surface methodology (RSM) is the widely used technique to optimise the process parameters. It is important for evaluating the influence of cutting parameters. This is a collection of statistical and mathematical concepts. With the help of RSM, it was reported that at high speed, the feed rate and point angle should be less in order to minimise the drilling-induced damage (Singh et al. 2008). The experimentation method is very expensive and takes time to complete the process. Therefore, the design of experiments (DOE) is used to reduce the unnecessary time associated with experiments.

10.7 CONCLUSION

The lightweight composite materials based on the natural fibres and polymers derived from renewable resources and their drilling characteristics are going to be a subject of major research concern in the future. The present chapter highlights the secondary manufacturing processes employed for polymer-based composite materials with emphasis on the current status and the future potential in conventional drilling of polymer composites. The fundamental challenges and the existing opportunities have also been discussed, which can help in formulating a roadmap for research and development of novel secondary manufacturing techniques. With the growing concerns of lean manufacturing and energy conservation, there is now a challenge for design engineers to develop technologies with less material wastage during machining operation. Formulation of machining problems and optimisation of process parameters using analysis of variance (ANOVA), response-surface methodology (RSM), Taguchi orthogonal array (DOE) and so on have minimised the experimental work for deciding the values of key factors associated with the drilling operation. In order to enlarge the application spectrum, still there is a need to develop new materials and processes to reduce the wastage in terms of material, energy and rejection.

REFERENCES

Ahmad. J. Y. S. 2009. Mechanics of chip formation. In *Machining of Polymer Composites*, Ahmad. J. Y. S. Ed. New York, USA: Springer Science and Business Media, pp. 63–109.
Bajpai, P. K., Debnath, K., and Singh, I. 2015. Hole making in natural fiber-reinforced polylactic acid laminates: An experimental investigation, *Journal of Thermoplastic Composite Materials* 30:30–46.
Bajpai, P. K. and Singh, I. 2013. Drilling behavior of sisal fiber-reinforced polypropylene composite laminates, *Journal of Reinforced Plastics and Composites* 32:1569–1576.

Bajpai, P. K., Singh, I., and Madaan, J. 2012. Joining of natural fiber reinforced composites using microwave energy: Experimental and finite element study, *Materials and Design* 35:596–602.

Debnath, K., Singh, I., and Dvivedi, A. 2014. Drilling characteristics of sisal fiber-reinforced epoxy and polypropylene composites, *Materials and Manufacturing Processes* 29:1401–1409.

Debnath, K., Singh, I., and Dvivedi, A. 2015a. On the analysis of force during secondary processing of natural fiber reinforced composite laminates, *Polymer Composites* 1–11. doi:10.1002/pc.23572.

Debnath, K., Singh, I., and Dvivedi, A. 2015b. Rotary mode ultrasonic drilling of glass fiber reinforced epoxy laminates, *Journal of Composite Materials* 49:949–963.

Holbery, J. and Houston, D. 2006. Natural-fiber-reinforced polymer composites in automotive applications, *Journal of The Minerals, Metals & Materials Society* 58:80–86.

Najaf, I. A., Kord, B., Abdi, A., and Ranaee, S. 2012. The impact of the nature of nano-clay on physical and mechanical properties of polypropylene/reed flour nanocomposites, *Journal of Thermoplastic Composite Materials* 25:717–727.

Nassar, M. M. A., Arunachalam, R., and Alzebdeh, K. I. 2016. Machinability of natural fiber reinforced composites: A review, *The International Journal of Advanced Manufacturing Technology* 1:1–20.

Rajakumar, I. P. T., Hariharan, P., and Vijayaraghavan, T. 2012. Drilling of carbon fibre reinforced plastic (CFRP) composites—A review, *International Journal of Materials and Product Technology* 43:43–67.

Rakesh, P. K., Singh, I., and Kumar, D. 2012. Drilling of composite laminates with solid and hollow drill point geometries, *Journal of Composite Materials* 46:3173–3180.

Singh, I. and Bajpai, P. 2013. Machining behavior of green composites: A comparison with conventional composites. In *Green Composites from Natural Resources*, Thakur, V. K., Ed. Boca Raton, FL: CRC Press/Taylor & Francis Group, pp. 267–280.

Singh, I., Bhatnagar, N., and Vishwanath, P. 2008. Drilling of uni-directional glass fiber reinforced plastics: Experimental and finite element study, *Materials and Design* 29:546–553.

Thomas, P. N. H. and Babistky, V. 2007. Experiments and simulations on ultrasonically assisted drilling, *Journal of Sound and Vibration* 308:815–830.

Zackrisson, L., Eriksson, I., and Backlund, J. 1994. Method and tool for machining a hole in a fiber-reinforced composite material, Swedish Patent No. 500933.

11 Evaluation of Surface Roughness in Turning with Precision Feed for Carbon Fibre-Reinforced Plastic Composites Using Response-Surface Methodology and Fuzzy Logic Modelling

K. Palanikumar and T. Rajasekaran

CONTENTS

11.1 INTRODUCTION

The major requirement in achieving quality of components in manufactured components is normally the geometrical precision and accuracy. This could be achieved by carrying out machining at micro-level in order to attain closer tolerances [1]. In the shop floor of any industry, use of precision machining is applied in various fields, which include aerospace, automotive, biomedical, electronics, environmental, communications, such as optical, bioengineering and photonics [2,3]. This is because they need micro-systems with the potential to ensure upgraded health care, growth in economy as well as high quality and long life of devices [4]. Machining of conventional materials such as metals has been investigated and analysed by many researchers. But machining of composite materials is not similar to that of machining of conventional materials such as metals. This is highly valid particularly in the case of polymeric materials and their composites due to their heterogeneity and anisotropic properties [5]. Even though carbon fibre-reinforced plastic (CFRP) composite materials are produced near to the net shape, final machining is often preferred to acquire the desirable surface quality in terms of dimensional accuracy [6]. Fibre-reinforced polymer (FRP) materials are finding numerous applications such as aircraft, automobiles, aesthetics [7–9] due to its excellent properties, namely high strength-to-weight ratio, high stiffness-to-weight ratio, high specific stiffness [10,11]. Identification of machining parameters at micro-level or sub-micro-level, which influences surface roughness for such a material after machining would be a definite contribution for manufacturing industry in achieving quality and performance of components [12,13].

Surface characteristic of machined material is generally believed to influence the performance of products significantly because they are directly connected to the capacity of material to experience stresses, friction, corrosion and temperature. There are many factors that bear responsibility for surface roughness of machined material such as machining parameters, machine tool vibration and cutting tool geometry, but machining parameters such as cutting speed, feed and depth of cut are believed to be controlled and offer a major contribution [14].

Few researchers including authors [15–17] have attempted to carry out various experimentations targeting higher accuracy by way of rough machining operation using macro-level range of machining parameters.

Brocks [18] tried to test the performance of resin transfer system for manufacturing composite parts. They analysed injection pressure with respect to fibre volume content, preform compression and permeability. Sene [19] prepared a nanocomposite based on epoxy and a mixture of two types of montmorillonite clays and multi-walled carbon nanotubes. They observed that low clay content did not significantly alter the studied flexural properties.

Assagra et al. [20] applied a composite coating based on geotextile fibres and polyurethane resin to overcome the drawback associated with single thin layer of resin that is suitable for wooden structures. Davim et al. [21] attempted to determine the machinability of PA66 polyamide using precision parameters. They used neat resin with 30% glass fibre reinforcement for carrying out turning operation using poly crystalline diamond (PCD) and ISO grade K15 (K15) cutting tools. It is evident from their findings that the addition of reinforcement affects the performance of cutting tool; radial force was higher than the cutting force and feed force; and further they observed that the tool materials and cutting conditions are responsible for lower forces.

Asao et al. [22] studied precision turning by means of a simplified predictive function of machining error such as predictive compensation for machining error, reliability of tool movement in computer numeric control (CNC) lathes. Wu [23] has carried out orthogonal micro-cutting according to strain gradient plasticity theory for modelling the material flow stress.

In this work, CFRP material in the form of solid rod is used to carry out turning operation using cemented carbide cutting tool insert. The machining parameters chosen are of micrometre level for carrying out experimentation. Then the outcome of the experiments, the surface roughness was measured using TR200 hand-held surface roughness measuring instrument. The surface roughness was modelled using response-surface methodology (RSM) and fuzzy logic and compared with the experimental value. The best methodology was identified based on the minimum error.

11.2 METHODOLOGY

Modelling of machining process is important for the industry. The model is developed to predict the response within the ranges of machining parameters studied. In the present investigation, two different methodologies are considered for predicting the machining parameters in machining of CFRP composites such as RSM and fuzzy logic modelling and are discussed in Sections 11.2.1 and 11.2.2.

11.2.1 RESPONSE-SURFACE METHODOLOGY

RSM is the combination of mathematical and statistical techniques used for the development of functional relationship between a response of interest, y and a number of associated control variables denoted by $x_1, x_2,...x_k$. The method is introduced in the early 1950s by Box and Wilson and this method is finding many applications in industries and research field including machining [24]. RSM uses quantitative data from appropriate experimental designs to determine and simultaneously to solve multivariate equations that are graphically represented as response surfaces, which can be used in three ways.

1. To establish an approximate relationship between y and $x_1, x_2,...x_k$, which can be used to predict the output response values for given settings of the selected control variables.
2. To determine, through hypothesis testing, significance of the factors whose levels are represented by $x_1, x_2,...x_k$.
3. To determine the optimum level settings of $x_1, x_2,...x_k$ that result in the maximizing or minimizing the response over a certain region of interest.

The relation between the machining parameters and the response surface roughness can be written in the linear form as follows:

$$\eta = \beta_0 + \beta_{1 \times 1} + \beta_{2 \times 2} + \beta_{3 \times 3} \tag{11.1}$$

Equation 11.1 could also be written in the following fashion:

$$\hat{y} = b_0 + b_{1\times1} + b_{2\times2} + b_{3\times3} \tag{11.2}$$

where:

$\hat{y}$ is the value of surface roughness by prediction with the transformation in logarithmic form

b_0, b_1, b_2 and b_3 are the coefficients of parameters β_0, β_1, β_2, β_3, respectively.

In the case of any statistical deficiency with the first-order model, the second-order model may be developed as follows:

$$y = \beta_0 + \sum_{i=1}^{k} \beta_i x_i + \sum \beta_{ii} x_i^2 + \sum_i \sum_{j \rangle i} \beta_{ij} x_i x_j + \varepsilon \tag{11.3}$$

The β coefficients, which should be determined in the second-order model, are obtained by the least square method [25].

The objective of this experimental study was to determine the effect of cutting parameters of micro-level on carbon fibre-reinforced polymer when carrying out turning operation using cemented carbide cutting tool.

11.2.2 Fuzzy Logic

A fuzzy intelligent system brings fuzzy sets and fuzzy logic to form one unit for knowledge representation and reasoning process [26]. It is one of the rapidly developing and most competing technologies applied to sophisticated systems [27]. Generally, the rule-based intelligent system is made up of a knowledge base or rule, a central database, an interpreter and meta rules to decide the validity of rule and to find out the order of rule searching, respectively, and makes inferences to unravel the problem. In fuzzy logic, four basic concepts such as fuzzy sets, linguistic variables, possibility distributions and fuzzy *if–then* rules, are used in most of the industrial applications [28]. Generally construction of rule base is done by using two types of fuzzy logic rules, either Mamdani type or Takagi-Sugeno-Kang (TSK)-type rules [29]. The act of judging the numerical values for machining operations is obtained from standard manuals or handbooks. The results from experiments conducted under certain environmental conditions cannot be directly used for non-conventional materials such as fibre-reinforced polymer composites.

In this work, selection of appropriate machining conditions using various membership functions of fuzzy logic for minimum surface roughness is done with three input variables and one output variable. Input variables are cutting speed, feed rate and depth of cut, and the output variable is surface roughness. Fuzzy logic uses membership functions, which is an arbitrary curve generally formed using straight lines and curves. The elementary stage in the fuzzy logic is the selection of appropriate shapes of the membership function for developing the algorithm in order to select the machining parameters.

11.3 SCHEME OF INVESTIGATION

CFRP rod was used for this experimental work. Unsaturated polyester resin with carbon fibre in the form of roving is reinforced by manual pultrusion with the unidirectional orientation of fibre. The rod was made with the diameter of 30 mm and with the length of 300 mm. Figure 11.1 illustrates the schematic picture of work piece material and tool with tool holder used for this study.

Cemented carbide cutting tools (K15) are found to be widely used in industries for the purpose of machining different categories of hard materials. This is due to the various mechanical properties; these tools possess high temperature strength, capability of sustaining its original shape even at elevated temperatures and higher hardness. The experimental setup used for turning contains a CNC lathe (ACE/Classic 2006) with the spindle speed of 4000 rpm, length of bed 500 mm, diameter of chuck 300 mm and eight number of turret tool head.

The surface roughness measured was the commonly used parameter, the arithmetic average surface roughness, R_a, which is the average of three values measured from three different locations on the circumference of the machined part to ensure reliability. TR200 hand-held surface roughness tester developed by Time Group Inc. equipped with a clear display of all measurement parameters and profile graphs on liquid-crystal display (LCD) monitor and Stylus Tip Radius of 2 μm was used for measuring surface roughness in this study. The surface roughness tester was set to a cut-off length of 0.8 mm. The measured profile is digitised and processed through the dedicated advanced surface finish analysis software Timesurf TR200 V1.0.

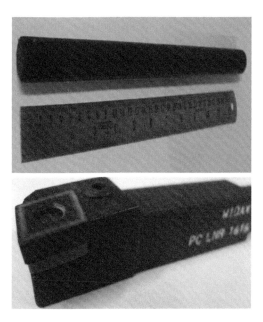

FIGURE 11.1 CFRP composite along with 1 ft scale and cutting tool along with tool holder.

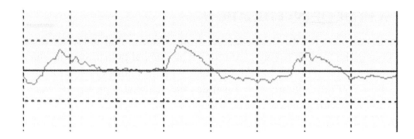

FIGURE 11.2   Typical surface profile obtained from machined surface.

The typical surface roughness profile obtained by tracing turned CFRP composite material using TR200 surface roughness instrument is shown in Figure 11.2.

It was planned to conduct experiment at atmospheric condition with 27 combinations between three cutting parameters, cutting speed, feed and depth of cut. It used Taguchi's orthogonal array of L_{27} (3^3) for easy conduction of experiment [30,31]. The machining conditions were used for machining composite material, and their levels are provided in Table 11.1. The machining parameters in terms of coded values, actual values and experimental values as well as predicted outcome surface roughness have been summarised in Table 11.2. The first three columns represent three machining parameters and their ranges in coded form. Another three columns show the same machining parameters in terms of actual values that are used for carrying out experimentation.

The scheme of the present investigation is presented in Figure 11.3.

TABLE 11.1

Machining Conditions and Their Levels Used for Machining Composite Material

S. No.	Cutting Speed (m/min)	Feed (mm/rev)	Depth of Cut (mm)
1	100	0.005	0.05
2	150	0.010	0.10
3	200	0.015	0.15

TABLE 11.2

Machining Parameters and Their Ranges along with the Response Obtained from Experimentation, RSM and Fuzzy Logic

S. No	Run Order	Coded Values			Actual Values			Surface Roughness, Ra (µm)		
		Cutting Speed	Feed	Depth of Cut	Cutting Speed (m/min)	Feed (mm/rev)	Depth of Cut (mm)	Experiment	RSM	Fuzzy
1	14	1	1	1	100	0.005	0.05	1.24	1.18	1.21
2	4	1	1	2	100	0.005	0.10	1.21	1.19	1.21
3	6	1	1	3	100	0.005	0.15	1.41	1.21	1.37
4	2	1	2	1	100	0.010	0.05	1.44	1.48	1.54
5	21	1	2	2	100	0.010	0.10	1.47	1.48	1.54
6	24	1	2	3	100	0.010	0.15	1.49	1.50	1.54
7	25	1	3	1	100	0.015	0.05	1.75	1.78	1.83
8	16	1	3	2	100	0.015	0.10	1.62	1.78	1.71
9	9	1	3	3	100	0.015	0.15	1.55	1.80	1.54
10	23	2	1	1	150	0.005	0.05	0.91	0.83	0.87
11	15	2	1	2	150	0.005	0.10	1.11	0.80	1.04
12	1	2	1	3	150	0.005	0.15	1.17	0.78	1.21
13	7	2	2	1	150	0.010	0.05	1.33	1.32	1.37
14	3	2	2	2	150	0.010	0.10	1.38	1.29	1.37
15	5	2	2	3	150	0.010	0.15	1.41	1.27	1.37
16	26	2	3	1	150	0.015	0.05	1.61	1.81	1.54
17	20	2	3	2	150	0.015	0.10	1.71	1.78	1.71
18	11	2	3	3	150	0.015	0.15	1.74	1.76	1.83
19	27	3	1	1	200	0.005	0.05	0.53	0.49	0.58
20	19	3	1	2	200	0.005	0.10	0.75	0.42	0.70
21	8	3	1	3	200	0.005	0.15	0.91	0.36	0.87
22	10	3	2	1	200	0.010	0.05	1.21	1.16	1.21
23	12	3	2	2	200	0.010	0.10	1.24	1.09	1.21
24	22	3	2	3	200	0.010	0.15	1.34	1.03	1.37
25	18	3	3	1	200	0.015	0.05	1.81	1.84	1.83
26	13	3	3	2	200	0.015	0.10	1.84	1.77	1.83
27	17	3	3	3	200	0.015	0.15	1.81	1.71	1.83

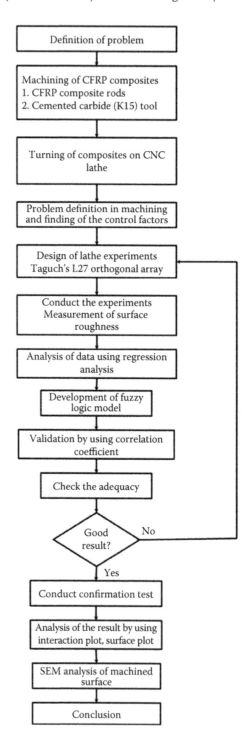

FIGURE 11.3   Scheme of the present investigation.

11.4 RESULTS AND DISCUSSION

Machining of composite material is an important requirement for the manufacturing industry. Machining cannot be avoided and is required for obtaining the near-net shape. The machining of CFRP composite materials is different from the conventional material. The results are analysed based on the modelling technique. Two different modelling techniques are adopted for analyzing the results:

- Response surface modelling
- Fuzzy logic modelling

11.4.1 RESPONSE SURFACE MODELLING

The analysis on influence of machining parameters on surface roughness is carried out using statistics. The empirical relation established for surface roughness in machining of CFRP composites is written as follows:

Surface roughness, R_a = 1.872 − 0.01013 cutting speed (m/min) − 14.9 feed (mm/rev.)

$$+ 1.59 \text{ depth of cut (mm)}$$

$$- 0.000003 \text{ cutting speed (m/min) * cutting speed (m/min)}$$

$$+ 133 \text{ feed (mm/rev.) * feed (mm/rev.)}$$

$$+ 0.00 \text{ depth of cut (mm) * depth of cut (mm)}$$

$$+ 0.7367 \text{ cutting speed (m/min) * feed (mm/rev.)}$$

$$+ 0.01633 \text{ cutting speed (m/min) * depth of cut (mm)}$$

$$- 293.3 \text{ feed (mm/rev.) * depth of cut (mm)}$$

The effectiveness of the model is tested by using coefficient of correlation $R\text{-}Sq$. The $R\text{-}Sq$ value of 0.9837 indicates that the model is effective. The adjusted $R\text{-}Sq$ values also calculated for the model, which is 0.9751. The $R\text{-}Sq$ and adjusted $R\text{-}Sq$ values are very close to each other and hence, the model is very effective for predicting the surface roughness in machining of CFRP composites. Further, the analysis of variance is carried out for the developed model. Table 11.3 shows the analysis of variance result for the surface roughness in machining CFRP composites. The analysis of variance analyses the influence of parameters, which affect the surface roughness in machining.

This table indicates the linear, square and interaction terms. From the analysis of the table, it has been known that the feed is the main important factor, which influences the surface roughness in machining of CFRP composites. The square effect does not show any effect on the surface roughness in machining. Among the interactions considered, the interaction between the parameters cutting speed and feed shows the very good effect. In machining of composite materials, feed is the

TABLE 11.3

Analysis of Variance for Surface Roughness

Source	Sum of Squares	DF	Mean Square	F-Value	Prob > F	
Model	2.851194	9	0.316799	114.088	<0.0001	Significant
Linear						
V – cutting speed (m/min)	0.16820	1	0.16820	60.57	<0.0001	
f – feed (mm/rev.)	2.13556	1	2.13556	769.07	< 0.0001	
d – depth of cut (mm)	0.05556	1	0.05556	20.01	0.0003	
Square						
V^2	0.000267	1	0.000267	0.096034	0.7604	
f^2	6.67E-05	1	6.67E-05	0.024008	0.8787	
d^2	0	1	0	0	1.0000	
Two-Way Interaction						
Vf	0.407008	1	0.407008	146.5747	<0.0001	
Vd	0.020008	1	0.020008	7.205543	0.0157	
fd	0.064533	1	0.064533	23.2402	0.0002	
Residual	0.047206	17	0.002777			
Corrected total	2.8984	26				

main factor which affects surface roughness. Here which is reflected in the cutting speed and feed interaction also. From the analysis of the table, the model is significant at 95% of confidence level. The model effectiveness (diagnosis) is evaluated by using the residual analysis. The plots obtained by analyzing the residuals are presented in Figure 11.4.

The residual plots give the relation between the developed model and error; it gives additional information about the model development process. The closeness of the developed model is indicated by the normal probability plots (Figure 11.4a). If almost all the points follow the straight line, then the model is said to be stable. Unwanted or abnormal distribution of the points indicates model inadequacy. In the present investigation, the points are following almost straight line, which indicates that the model is effective. Figure 11.4b shows the relation between the externally studentised residuals and the predicted values; the results indicate that the residuals are distributed within the limits and hence the model is said to be adequate. Figure 11.4c shows the relation between the residuals and run number. The result shows that the values are distributed in both the positive and negative directions, which show that the experimental results are distributed evenly. The relation between the predicted and actual results is presented in Figure 11.4d. The results show the close relationship between the developed model and experimental values. As a whole, the diagnosis results do not show any model inadequacy and hence the developed models are very effective to predict the surface roughness in machining of CFRP composites.

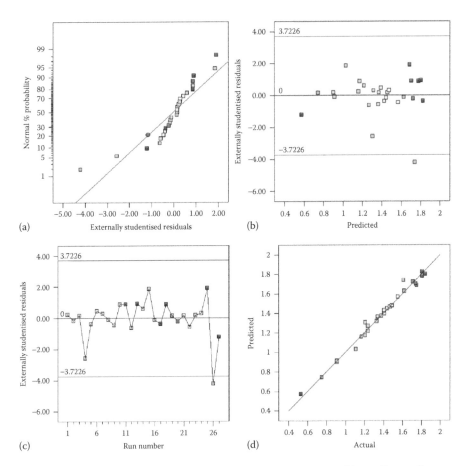

FIGURE 11.4 Diagnosis checking of the model (a) Normal probability vs Externally studentised residuals, (b) Predicted results vs Externally studentised residuals, (c) Run number vs Externally studentised residuals, and (d) Actual vs Predicted results.

11.4.2 Fuzzy Logic Modelling

Fuzzy logic supports 11 types of membership functions. Although there are many number of membership functions available such as triangular, trapezoidal, Gaussian and so on, the modelling of the surface roughness in this chapter is using the triangular type as described in Reference 32. The membership functions for the input machining parameters such as cutting speed, feed and depth of cut have been divided into three sets, namely low, medium and high. Triangule-shaped membership functions for the aforementioned machining parameters have been shown in Figure 11.5 through 11.7, respectively.

Output machining parameter surface roughness has been divided into nine sets such as lowest, lower, low, low medium, medium, high medium, high, higher and highest. Fuzzy logic uses membership functions, which is an arbitrary curve.

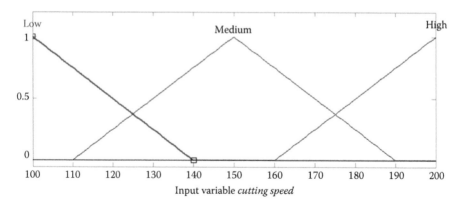

FIGURE 11.5   Membership functions for cutting speed.

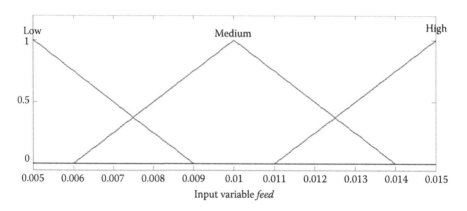

FIGURE 11.6   Membership functions for feed.

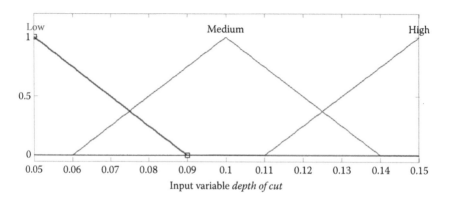

FIGURE 11.7   Membership functions for depth of cut.

Triangule-type membership functions are simpler and easy to use than the other membership functions. Triangle-type membership function is specified by three parameters, but others need more parameters. It needs simple formula, and their computational efficiency is good and is extensively used in real-time implementations and hence, triangular membership functions are used in this work [33]. The triangule-shaped membership function for input is specified by three parameters {a, b, c} as follows:

Triangle x; a,b,c

$$
\text{Triangle }(x; a, b, c) = \begin{cases} 0, & x \le a \\ \dfrac{x-a}{b-a}, & a \le x \le b \\ \dfrac{c-x}{c-b}, & b \le x \le c \\ 0, & c \le x \end{cases}
\tag{11.4}
$$

where a, b, c stands for the triangular fuzzy triplet, and it determines the x coordinates of the three corners underlying triangular membership function [34]. Similarly triangule-shaped membership function for surface roughness has been shown in Figure 11.8.

The concept of fuzzy reasoning for three-input one-output fuzzy logic unit is described as follows:

The fuzzy rule base consists of a group of *if–then* statements with three inputs, x_1, x_2 and x_3 and one output, y [35]. By taking the max–min compositional operation, the fuzzy reasoning of these rules yields a fuzzy output:

$$
\mu_{D_o}(y) = [\mu_{A_1}(x_1) \wedge \mu_{B_1}(x_2) \wedge \mu_{C_1}(x_3) \wedge \mu_{D_1}(y) \vee \ldots
$$
$$
\mu_{A_n}(x_1) \wedge \mu_{B_n}(x_2) \wedge \mu_{C_n}(x_3) \wedge \mu_{D_n}(y)]
\tag{11.5}
$$

Finally, a defuzzification method is used to transform the fuzzy output into a nonfuzzy value, y_0.

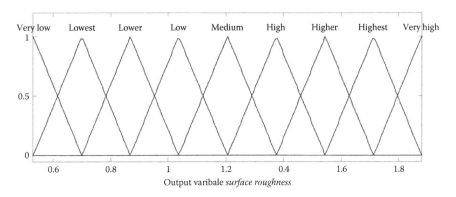

FIGURE 11.8 Membership functions for surface roughness.

Defuzzification is carried out using centroid defuzzification method. It produces the centre area of the possibility distribution of the inferenced output. It is also one of the more frequently used defuzzification method that calculates the centroid of the area under the membership function:

$$y_0 = \frac{\sum y\mu_{D_0}(y)}{\sum \mu_{D_0}(y)} \tag{11.6}$$

The non-fuzzy value y_0 gives the output value in numerical form. The result obtained from using the fuzzy logic model is presented in Table 11.2.

Regression statistics $R\text{-}Sq$ obtained is 95.9%, which indicates that the machining parameters show 95.9% of variance in surface roughness and an evidence that the presented model fits closer to the experimental value.

11.4.3 Analysis of the Results

Statistical analysis was performed with the help of Minitab® software for obtaining main effects. The main effect of a factor is described as the average change in the response generated by the change in the level of input factors studied. In case, if the line is horizontal in the main effects plot, then there is no main effect. The magnitude of the main effect depends upon the difference in the vertical position of the plotted points. When the vertical position of plotted points is found to be with a larger difference, the magnitude of the main effect would also have much difference [36].

Figure 11.9 shows the response graph for the given experimental results on surface roughness, which has been generated by the software Minitab®. The response graph for surface roughness has been developed by way of grouping the response obtained from experimentation. This is done by averaging the responses of each

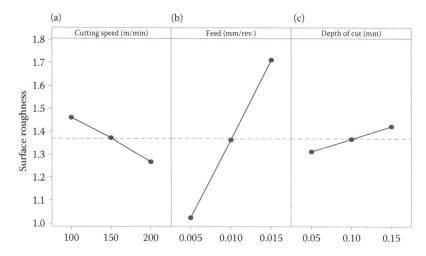

FIGURE 11.9 (a–c) Response graph for surface roughness.

level corresponding to the different parameters. Greater the length of the curve or line, higher the rank, higher the predominant level the parameter possesses. This trend indicates that the feed has a strong influence on the surface roughness characteristics compared with the other machining parameters subsequently followed by other parameters such as cutting speed and depth of cut.

The trend of machining parameters on surface roughness is presented in Figure 11.9a–c. On referring Figure 11.9a, it is observed that the value of surface roughness is decreasing with the increase in the value of cutting speed. But interpreting at another part of Figure 11.9b for the trend of cutting speed, it is observed that the value of surface roughness is on increasing trend with the increase of feed. Figure 11.9c refers the trend on surface roughness with respect to depth of cut, which does have strong influence on the response parameter.

Although the parameter depth of cut does not have any significant role on the part of the surface roughness, its trend was observed that the surface roughness was increasing on the increasing value of depth of cut. Therefore, the combination of machining parameter that results in better surface roughness could be lower range of feed along with the higher ranges of cutting speed accompanied by lower value of depth of cut.

Further, three-dimensional response surface plots for various machining parameters for surface roughness are also shown in Figures 11.10 through 11.12. These plots show how a response variable relates to two factors based on a model equation while any other factors are held constant. These plots are useful for establishing desirable response values and for operating conditions.

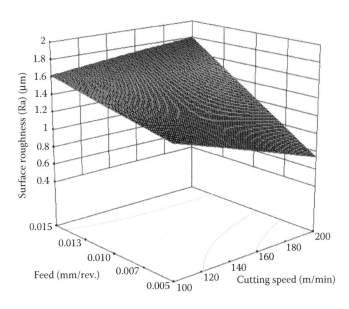

FIGURE 11.10 Three-dimensional response graph for surface roughness by varying the parameters cutting speed and feed.

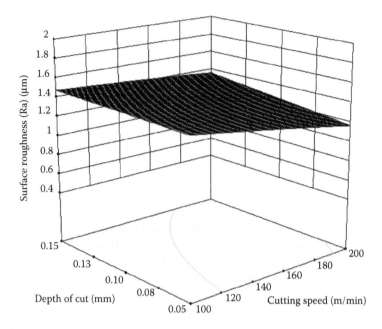

FIGURE 11.11   Three-dimensional response graph for surface roughness by varying the parameters cutting speed and depth of cut.

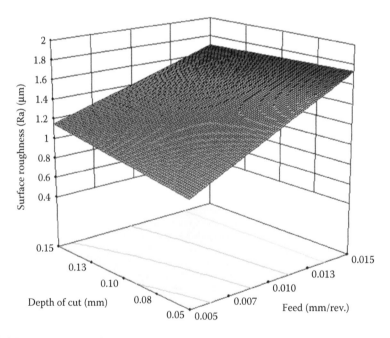

FIGURE 11.12   Three-dimensional response graph for surface roughness by varying the parameters feed and depth of cut.

Figure 11.10 demonstrates the trend between cutting speed and feed while keeping depth of cut at a constant level of 0.1 mm. It is observed from the plot that the surface roughness increases linearly with respect to the feed and also highly sensitive to the feed, coinciding with the result that is obtained from the aforementioned response table. This is attributed to the fact that the increase in feed increases the contact area between the workpiece and the cutting tool, resulting in the generation of more feed force. This produces more pulling and fracture of fibres, yielding higher surface roughness. The parameter feed is ranked as 1 that the feed is the prime role on surface roughness. To minimise surface roughness, it would be better to choose settings for cutting speed and feed in the bottom of the right side of the plot, specifying higher cutting speed and lower value of feed. The trend of cutting speed versus depth of cut maintaining a constant feed of 0.01 mm/rev. is shown in Figure 11.11. Similarly, the darkest area indicates the lowest surface roughness. It is observed that increasing feed increases the surface roughness linearly at the lower range and non-linearly at the higher level. It is also observed that increasing cutting speed decreases surface roughness, indicating that a combination of lower feed and higher cutting speed could yield a better surface roughness in turning CFRP composites.

In Figure 11.12, the surface plot shows the characteristics of feed versus depth of cut with the cutting speed level of 100 m/min. This plot shows how feed and depth of cut influence the response surface roughness when cutting speed is kept at its middle level.

Interaction graph on surface roughness for various machining parameters such as cutting speed, feed and depth of cut is illustrated in Figure 11.13. It is observed from

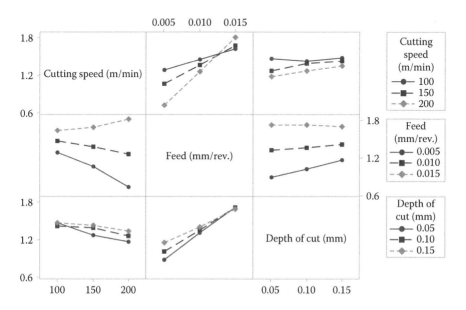

FIGURE 11.13   Interaction plot for surface roughness.

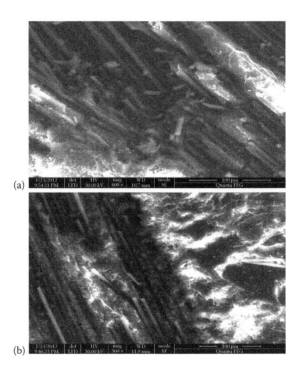

FIGURE 11.14　SEM images of CFRP composites (a) at $f = 0.005$ and (b) $f = 0.015$ mm/rev.

the graph that there is a good interaction between the parameters feed and cutting speed as the trend lines are not parallel to each other for these parameters.

Figure 11.14a and b illustrates scanning electron microscope (SEM) images of CFRP composite after machining, using different machining conditions. It is observed from the figure that there exists good bonding between resin and fibre. Figure 11.14a that corresponds to the condition $f = 0.005$ mm/rev. shows only a few fractures of fibre attributing low surface roughness, because surface roughness depends on the quantity of fibre fracture on the surface of the machined composites. Similarly, Figure 11.14b that corresponds to the feed condition $f = 0.015$ mm/rev. shows full of fractured fibres attributed to the higher surface roughness. From the results, it has been asserted that the high speed, low feed and moderate depth of cut are preferred for turning of composite materials.

The results obtained from conducting machining experiment as well as those obtained from response surface modelling and fuzzy modelling are compared for identifying its correlation with each other that is well shown in Figure 11.15.

It is observed from the figure that among the two methods, the predicted values that are obtained from fuzzy logic modelling and response surface modeling both were in good correlation for the range of machining parameters chosen for the present work. Therefore, both modelling technique could be successfully used for prediction of surface roughness in machining composite material.

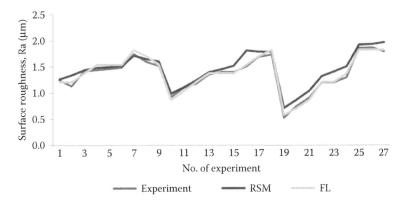

FIGURE 11.15 Comparison of surface roughness from experiments, RSM and fuzzy logic.

11.4.4 CONFIRMATION EXPERIMENT

The results obtained from the developed model and experimental investigation are verified using the confirmation experiments. The confirmation experimental results are carried out at three selected feed cutting conditions, maintaining cutting speed and depth of cut constant and are presented in Figure 11.16. The results revealed that the experimental results are very close to the predicted results and hence, the developed models can be effectively used for the prediction of surface roughness in precision turning of CFRP composite material.

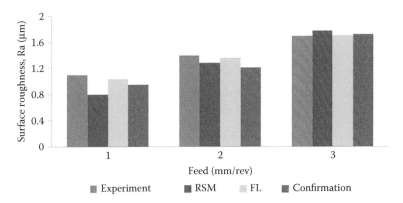

FIGURE 11.16 Comparison of surface roughness from experiments, RSM, fuzzy logic and confirmation trial.

11.5   SUMMARY

Based on the experiments conducted and observations made from this work, the following concluding remarks may be made:

- Use of precision-level machining parameters resulted in a relatively better surface roughness in turning of CFRP composite materials that is ranging from 0.53 to 1.88 μm.
- The machining parameter that leads to influencing the surface roughness is feed followed by cutting speed and depth of cut. As far as the trend of machining parameters is concerned, surface roughness increased with the increase in feed, but it was found to be better with the increase in cutting speed.
- Response surface and fuzzy methodologies were used for modelling response, the surface roughness. Both methods were producing lowest percentage error with the experimental values.
- The combination of machining parameters that offer the desired surface roughness is low value of feed accompanied by higher cutting speed.

REFERENCES

1. Weck M, Hartel R. Design, manufacture and testing of precision machines with essential polymer concrete components. *Precision Engineering* 1985; 7(3): 165–170.
2. Wang P, Zhang X, Lim G, Neo H, Malcolm AA, Xiang Y, Lu G, Yang J. Improvement of impact-resistant property of glass fiber reinforced composites by carbon nanotube-modified epoxy and pre-stretched fiber fabrics. *Journal of Materials Science* 2015; 50: 5978–5992.
3. Mamalis AG, Lavrynenko SN. On the precision single-point diamond machining of polymeric materials. *Journal of Materials Processing Technology* 2007; 181: 203–205.
4. Chae J, Park SS, Freiheit T. Investigation of micro-cutting operations. *International Journal of Machine Tools & Manufacture* 2006; 46: 313–332.
5. Wang H, Sun J, Li J, Lu L, Li N. Evaluation of cutting force and cutting temperature in milling carbon fiber-reinforced polymer composites. *International Journal of Advanced Manufacturing Technology* 2016; 82: 1517–1525. doi:10.1007/s00170-015-7479-2.
6. Palanikumar K. Modeling and analysis for surface roughness in machining glass fibre reinforced plastics using response surface methodology. *Materials and Design* 2007; 28: 2611–2618.
7. Palanikumar K, Mata F, Paulo Davim J. Analysis of surface roughness parameters in turning of FRP tubes by PCD tool. *Journal of Materials Processing Technology* 2008; 204: 469–474.
8. Palanikumar K, Campos Rubio J, Abrao AM, Esteves Corriea A, Paulo Davim J. Influence of drill point angle in high speed drilling of glass fiber reinforced plastics. *Journal of Composite Materials* 2008; 42: 2585–2597. doi:10.1177/0021998308096322.
9. Palanikumar K, Latha B, Senthilkumar VS, Karunamurthy R. Multiple performance optimization in machining GFRP composites by a PCD tool using Non dominated sorting GA (NSGA II). *Metals and Materials International* 2009; 15(2): 249–258.
10. Himmel N, Pfaff T, Schmitt U, Glitza KW. Development of high performance CFRP shell structures for the pixel detected in the ATLAS experiment at CERN. *Composites Science and Technology* 2003; 63: 2013–2028.

11. Arumugam V, Sidharth AAP, Santuli C. Characterization of failure modes in compression after impact of glass-epoxy composite laminates using acoustic emission monitoring. *Journal of the Brazilian Society of Mechanical Sciences and Engineering* 2015; 37: 1445–1455. doi:10.1007/s40430-015-0263-7.

12. Malhotra SK, Ganesan N, Veluswami MA. Effect of fiber orientation and boundary conditions on the vibration behavior of orthotropic square plates. *Composite Structures* 1988; 9: 247–255.

13. Liu K, Melkote SN. Effect of plastic side flow on surface roughness in microturning process. *International Journal of Machine Tools & Manufacture* 2006; 46: 1778–1785.

14. Hagiwara M, Chen S, Jawahir IS. Contour finish turning operations with coated grooved tools. Optimization of machining performance. *Journal of Materials Processing Technology* 2009; 209: 332–342.

15. Rajasekaran T, Palanikumar K, Vinayagam BK. Application of fuzzy logic for modeling surface roughness in turning CFRP using CBN tool. *Production Engineering Research and Development* 2011; 5(2): 191–199.

16. Rajasekaran T, Palanikumar K, Vinayagam BK. Experimental investigation and analysis in turning of CFRP composites. *Journal of Composite Materials* 2011; 46(7): 809–821.

17. Palanikumar K, Rajasekaran T, Latha B. Fuzzy rule-based modeling of machining parameters for surface roughness in turning carbon particle-reinforced polyamide. *Journal of Thermoplastic Composite Materials* 2015; 28(10): 1387–1405.

18. Brocks T, Shiino MY, Cioffi MOH, Voorwald HJC, Filho AC. Experimental RTM manufacturing analysis of carbon/epoxy composites for aerospace application: Non-crimp and woven fabric differences. *Materials Research* 2013; 16(5): 1175–1182.

19. Sene TS, da Silva LV, Amico SC, Beckera D, Ramirez AM, Coelho LAF. Glass fiber hybrid composites molded by RTM using a dispersion of carbon nanotubes/clay in epoxy. *Materials Research* 2013; 16(5): 1128–1133.

20. Assagra YAO, Altafim RAP, da Silva JFR, Basso HC, Lahr FAR, Altafim RAC. Laminated composite based on polyester geotextile fibers and polyurethane resin for coating wood structures. *Materials Research* 2013; 16(5): 1140–1147.

21. Davim JP, Silva LR, Festas A, Abrão AM. Machinability study on precision turning of PA66 polyamide with and without glass fibre reinforcing. *Materials and Design* 2009; 30: 228–234. doi:10.1016/j.matdes.2008.05.003.

22. Asao T, Mizugaki Y, Sakamoto M. Precision turning by means of a simplified. Predictive function of machining error. *Annals of the CIRP* 1992; 41(1): 447–450.

23. Wu J, Liu Z. Modeling of flow stress in orthogonal micro cutting process based on strain gradient plasticity theory. *International Journal of Advanced Manufacturing Technology* 2010; 46: 143–149.

24. Box GEP, Wilson KB. On the experimental attainment of optimum conditions (with discussion). *Journal of the Royal Statistical Society Series* 1951; B13(1): 1–45.

25. Palanikumar K. Modeling and analysis for surface roughness in machining glass fibre reinforced plastics using response surface methodology. *Materials and Design* 2007; 28: 2611–2618.

26. Akay M, Cohen M, Hudson D. Fuzzy sets in life sciences. *Fuzzy Sets and Systems* 1997; 90: 219–224.

27. Yen J, Langari R. *Fuzzy Logic Intelligence, Control, and Information.* Dorling Kindersley (India) Pvt. Ltd., Noida, India, 2006.

28. Nandi AK, Davim JP. A study of drilling performances with minimum quantity of lubricant using fuzzy logic rules. *Mechatronics* 2008; 19: 218–232. doi:10.1016/j.mechatronics.2008.08.004.

29. Peres CR, Guerra REH, Haber RH, Alique A, Ros S. Fuzzy model and hierarchical fuzzy control integration: an approach for milling process optimization. *Computers in Industry* 1999; 39: 199–207.

30. Lin JL, Wang KS, Yan BH, Tarng YS. Optimization of the electrical discharge machining process based on the Taguchi method with fuzzy logic. *Journal of Materials Processing Technology* 2000; 102: 48–55.

31. Tzeng CJ, Lin YH, Yang YK, Jeng MC. Optimization of turning operations with multiple performance characteristics using the Taguchi method and grey relational analysis. *Journal of Materials Processing Technology* 2009; 209: 2753–2759.

32. Sharif Ullah AMM, Harib KH. A human-assisted knowledge extraction method for machining operations. *Advanced Engineering Informatics* 2006; 20: 335–350.

33. Jang JSR, Sun CT, Mizutani E. *Neuro-Fuzzy and Soft Computing—A Computational Approach to Learning and Machine Intelligence.* Pearson Education, Pvt Ltd, Singapore, 2004, pp. 24–25999.

34. Rajasekaran T, Palanikumar K, Vinayagam BK. Application of fuzzy logic for modeling surface roughness, in turning CFRP composites using CBN tool. *Production Engineering Research and Development* 2011; 5: 191–199.

35. Latha B, Senthilkumar VS. Analysis of thrust force in drilling glass fiber-reinforced plastic composites using fuzzy logic. *Materials and Manufacturing Processes* 24(4): 509–516.

36. Tiryaki S, Coşkun Hamzaçebi, Abdulkadir Malkoçoğlu. Evaluation of process parameters for lower surface roughness in wood machining by using Taguchi design methodology. *European Journal of Wood and Wood Products* 2015; 73: 537–545.

12 Research Progress in the Area of Advanced Machining of Polymer Matrix Composites

Kishore Debnath, M. Roy Choudhury and Jung Il Song

CONTENTS

12.1 INTRODUCTION

The conventional machining of polymer composites results in significant damage to the machined surface in the form of delamination, fibre pull-out, splintering, spalling and poor surface finish. These types of damages occur due to the physical interaction between the tool and the workpiece during machining [1]. Therefore, the advanced machining techniques have drawn much attention of many researchers as there is no direct interaction between the tool and workpiece during machining [2,3]. The advanced machining techniques can be used as a potential alternative to the conventional machining methods because advanced machining of polymer composites results in less damage to the machined surface as cutting forces are not generated during the machining operation. The advanced machining techniques are practised when the primary requirements are high accuracy and precision. Unlike the conventional machining of polymer composites, the mechanical and thermal damages are less in the case of advanced machining of polymer composites. The negligible heat-affected zone, reduced waste materials, better surface quality, noise- and dust-free machining and process flexibility are some additional advantages of the advanced machining techniques. Moreover, the advanced machining techniques can be used to machine a wide variety of composites with better surface quality. The subsequent sections present and discuss the metal removal mechanism of differ- ent advanced machining techniques such as electric discharge machining, electro- chemical machining, water jet machining, abrasive water jet machining, laser beam machining and vibration-assisted machining in the context of polymer composites. The influence of various machining parameters on the quality of the machined sur- face has been reviewed. The challenges with the various advanced machining tech- niques have also been adequately discussed.

12.2 ELECTRIC DISCHARGE MACHINING OF POLYMER COMPOSITES

In electric discharge machining, the material is removed due to the melting and vapourising of the work material. A series of electric sparks is generated by estab- lishing a potential difference between the electrode and workpiece. The bombard- ment of electric spark results in thermal erosion of both work and tool materials and thus, causes the removal of work material and wear of tool, respectively. The schematic of a typical electric discharge machining setup is shown in Figure 12.1.

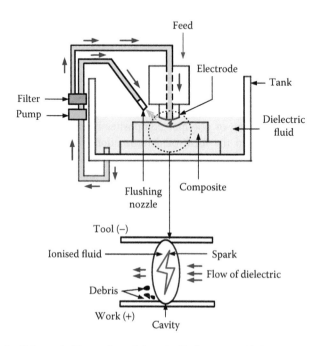

FIGURE 12.1   Schematic illustration of electric discharge machining.

The main limitation of the electric discharge machining is that it is not suitable for electrically non-conductive materials. The electrical conductivity of the material should be more than $200/\Omega/cm$ to initiate the electric sparks between the workpiece and tool. Therefore, polymer composites are not a suitable candidate for electric discharge machining as these materials have low electrical conductivity. However, the electrical conductivity of the polymer composites can be improved by incorporating metallic fillers such as copper, aluminium and silver powders. The electric discharge machining process can be potentially applied for machining of carbon fibre-reinforced composite as the electrical conductivity of the carbon fibres is much higher as compared to the other reinforcing fibres such as glass and aramid fibre. Nakamura et al. [4] investigated electric discharge machining of ceramic composites. Konig et al. [5] reported that material removal rate during electric discharge machining of carbon fibre-reinforced composites is higher than that of ceramic materials. The conductive nature of graphite fibres makes the composite fairly suitable for electric discharge machining. Lau et al. [6] investigated the effect of two different electrode materials on the machining performance of carbon fibre-reinforced epoxy laminates. It was found that the copper electrode wears out slowly as compared to the graphite electrode. The experimental findings also disclosed that the material removal rate increases and wear ratio decreases with positive polarity of the electrode. The damages such as isolated craters, thermal expansion, distortion of fibres, cracking of matrix interface and delamination are formed during electric discharge machining of carbon fibre-reinforced epoxy laminates. Guu et al. [7] revealed that the extent of delamination, surface roughness and recast layer thickness increases with an increase in input power during electric discharge machining of the carbon–carbon composites. It was also concluded that a smaller thermal gradient develops at a lower pulse current, whereas a higher pulse current results in the development of a steeper thermal gradient. The steeper thermal gradient results in the formation of recast layer on the machined surface. The formation of recast layer in turn results in low surface finish of the machined components. However, the holes produced during electric discharge machining are free from burr and are of superior quality than that of conventional drilling. George et al. [8] investigated the influence of pulse current, pulse on-time and gap voltage on the material removal rate and electrode wear rate during electric discharge machining of carbon–carbon composites. It was found that pulse on-time has negligible influence on the output responses among all input parameters.

12.3   ELECTROCHEMICAL MACHINING
OF POLYMER COMPOSITES

The main advantage of electrochemical machining over electric discharge machining is that it is feasible for machining of both electrically conductive and non-conductive materials, regardless of their inherent properties [9–12]. The machined surface produced by the electrochemical machining is exceptionally smooth. In this process, an electrolytic cell is formed by the cathode tool and an anode workpiece with a suitable electrolyte flowing between them. The anode

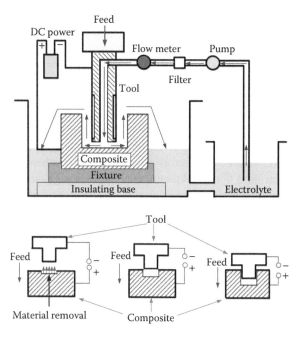

FIGURE 12.2   Schematic illustration of electrochemical machining.

workpiece is dissolved according to Faraday's law when a sufficient voltage is applied across the gap between the anode and the cathode. In this process, the material is removed due to the melting, vaporisation and mechanical abrasion due to cavitation of the formed gas bubbles. Figure 12.2 shows the schematic of a typical electrochemical machining setup. Tandon et al. [9] experimentally investigated the feasibility of electrochemical spark machining in cutting and drilling of glass and Kevlar fibre-reinforced epoxy composites. The authors have reported that material removal rate, tool wear rate, relative tool wear and diametral overcut increase with an increase in the specific conductivity of the electrolyte but decrease with an increase in the diameter of the tool during electrochemical spark drilling of Kevlar fibre-reinforced epoxy composites. On the other hand, the material removal rate, tool wear rate and overcut were found to be increased, and relative tool wear was found to be decreased with an increase in the applied voltage across the electrodes and in the specific conductivity of the electrolyte during electrochemical spark cutting of glass fibre-reinforced epoxy composites. Abrate and Walton [11] reported that electrochemical reactions do not take place for electrically non-conductive materials. The electrochemical action is involved only in the generation of the bubbles. Jain et al. [13,14] performed electrochemical spark machining of Kevlar and glass fibre-reinforced epoxy composites taking copper as an electrode. The electrolytes used were an aqueous solution of sodium hydroxide and sodium chloride. It was concluded that electrochemical spark machining is a viable solution for cutting of fibre-reinforced polymer composites. The authors have identified that the thermomechanical phenomena were the main source of material removal in electrochemical spark machining of polymer composites.

12.4 WATER JET MACHINING AND ABRASIVE WATER JET MACHINING OF POLYMER COMPOSITES

The water jet machining is capable of machining a wide range of composites by utilising the impact energy of pressurised jet water [15]. In water jet machining, a high-pressure and high-velocity jet of water exert a high impact on the workpiece, which results in the removal of material from the chosen workpiece. The materials may remove due to the shearing, micro-machining and erosion. The failures such as micro-bending-induced fracture and out-of-plane shear have been observed during water jet machining of composites. The water jet machining can be utilised for cutting, milling and drilling of polymer composites. When high-pressure water jet in addition with fine abrasive particles is used for machining, it is called abrasive water jet machining. The abrasive water jet machining is a modern technique, which can machine inhomogeneous, hard and abrasive materials. A typical abrasive water jet machining setup is shown in Figure 12.3. The major advantages of abrasive water jet machining include negligible heat-affected zone, higher machining speed and impose nominal stresses on the work material [16].

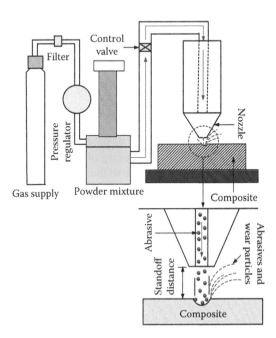

FIGURE 12.3 Schematic illustration of abrasive water jet machining.

The narrow kerf and high machining versatility and flexibility are the additional characteristics of this advanced machining technique [17]. Moreover, the process is the most environmentally friendly machining process as the process does not require any cutting fluids and generates no fumes or toxic gasses and harmful substances, unlike conventional machining processes. But the major disadvantages of the abrasive water jet machining include (1) low material removal rate, (2) higher noise, (3) inability to machine blind holes and (4) damage to the machine parts due to the action of abrasives. Some of the polymer composites suitable for abrasive water jet machining are boron fibre-reinforced epoxy composites, boron fibre-reinforced polyester composites, glass fibre-reinforced epoxy composites, graphite fibre-reinforced epoxy composites and aramid fibre-reinforced epoxy composites. Ramulu and Arola [18] reported that the abrasive water jet machining is more suitable than the water jet machining in terms of higher material removal rate and superior surface finish during machining of unidirectional graphite fibre-reinforced epoxy composite. Hashish [19] investigated the abrasive water jet machining of metal matrix composites, laminated thermoset composites and ceramic composites. The machining operations considered are linear cutting, turning, milling and drilling of small-diameter holes. It was reported that the three basic steps involved in abrasive water jet drilling of advanced composites are (1) piercing, (2) kerf cutting and (3) milling. The range of water pressure and jet diameter used during water jet machining of composites are 200–400 MPa and 0.08–0.5 mm, respectively [20]. Kim et al. [21] reported that the penetration rate of the abrasive particle decreases with an increase in the depth of the hole because the return flow of the jet after striking the laminate may reduce the velocity of the abrasive particle. The water pressure requires for machining brittle materials such as glass is much lower than that of ceramics. The materials with low yield strength can be machined with straight water jet machining, that is, without the aid of abrasives. Shanmugam and co-workers [22] investigated the delamination produced during water jet machining of graphite fibre-reinforced epoxy composites. It was reported that the delamination is initiated due to the impact of a high pressurised jet of water. The cracks are formed on the surface of the laminate due to the impact of the high pressurised jet resulting in delamination of the composite. The shock loading of water and hydrodynamic pressurisation are also responsible for the damage to the machined surface. The delamination was found to be predominant at a higher traverse speed of water jet and at a lower flow rate of abrasives [23]. Hashish [19] reported that damage induced during water jet machining can be reduced by lowering the water pressure and by reducing the jet size. On the other hand, the delamination can be reduced when machining is performed at a higher standoff distance. Hamatani and Ramulu [24] found that the damage is induced at the top of the laminate due to the high impact of abrasives, which can be eliminated by decreasing the standoff distance. But at lower standoff distance, the kerf taper ratio decreases. Hashish [25] developed a technique to produce high-quality and small-diameter holes during abrasive water jet machining of composites. The shape, quality and diameter of the hole can be controlled by adjusting the jet pressure, time and abrasive flow rate. The high accuracy hole can be produced during drilling of ceramic matrix composite using this technique. The water jet machining of glass fibre-reinforced epoxy composites revealed that the cutting orientation, standoff distance and abrasive mass flow rate are the insignificant parameters in controlling the surface roughness [26]. Azmir and Ahsan [27] investigated the responses of various factors, namely abrasive type, hydraulic pressure, standoff distance, abrasive flow rate, traverse rate and cutting orientation on the surface roughness and kerf taper ratio during abrasive water jet machining of glass fibre-reinforced epoxy composites. The statistical analysis showed that the jet pressure and abrasive materials chosen for the purpose of investigation were the significant factors influencing the response parameters. It was also observed that the machining performance was better when jet pressure and abrasives flow rate were high. It was also reported that the high traverse speed of water jet was the most effective parameter, and the mass flow rate was the less effective parameter on characterising the surface roughness and the taper ratio during abrasive water jet machining of Kevlar fibre-reinforced phenolic composites.

12.5 LASER BEAM MACHINING OF POLYMER COMPOSITES

The basic scheme of laser beam machining process is shown in Figure 12.4. The laser beam machining can produce high-quality holes by focusing intense and highly directional coherent laser beam on the workpiece to be drilled. The holes are generated due to melting and vaporisation of the work material or sometimes

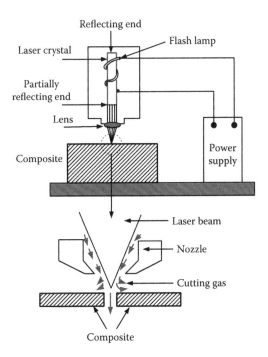

FIGURE 12.4 Schematic illustration of laser beam machining.

due to the chemical degradation of work material. A high-energy infrared beam of the diameter of 0.1–1 mm is spotted on the target workpiece. A jet of gas coaxial with the laser beam is focussed on the work surface to remove the fluids and degradation products from the machining zone. Generally, carbon dioxide (CO_2) and helium are used as assist gasses during laser beam machining of glass and boron fibre-reinforced composites, respectively [28]. Air is used as assist gas for machining of non-metallic materials [29]. The problems associated with chatter and vibration during conventional machining is significantly eliminated in laser beam machining as it is a *contact-free* process. Moreover, thin components can also be machined using laser beam technique [30]. The laser beam machining is suitable over water jet machining and abrasive water jet machining of polymer composites due to its specific characteristics such as low operating cost, free of noise and generation of small kerf. The laser beam machining is also suitable for machining of plastic materials as plastic possesses higher absorption coefficient for infrared radiation and lower thermal conductivity [31]. The thermosetting plastics (epoxy resin) require relatively high temperature and energy for melting and shearing. CO_2 and neodymium-doped yttrium aluminium garnet (Nd:YAG) lasers are two widely used laser sources for the purpose of machining of materials [32]. The radiations emitted by the CO_2 laser can be easily absorbed by non-metals and polymer composites, but Nd:YAG laser cannot sufficiently be absorbed by the organic matrices. It was found that Nd:YAG laser is more suitable than the CO_2 laser for drilling of carbon fibre-reinforced composites as it can significantly reduce the damage associated with machining [33,34]. It was also stated that the aramid fibre-reinforced composites can be easily machined using laser beam followed by glass and carbon fibre-reinforced composite [35,36]. Young [37] investigated the fatigue behaviour during Nd:YAG laser drilling of carbon fibre-reinforced composites. It was observed that the damage formed in and around the drilled hole during laser drilling of carbon fibre-reinforced thermoplastic polyetheretherketone (PEEK) composites was less as compared to the laser drilling of carbon fibre-reinforced thermosetting (epoxy) composites. The machining characteristics of laser beam drilling depend on the energy absorption capacity and thermal diffusivity of the material. Yung et al. [38] found that the heat-affected zone produced during ultraviolet-yttrium aluminium garnet (UV-YAG) laser drilling of glass fibre-reinforced composites depends on the average laser power and pulse repetition rate. At low laser power and low pulse repetition rate, hole produced by laser beam drilling was quite clean, whereas vaporisation of epoxy resin and melting of glass fibres were observed at high laser power and high pulse repetition rate.

12.6 VIBRATION-ASSISTED MACHINING OF POLYMER COMPOSITES

The vibration-assisted drilling is an innovative drilling method that has been developed to meet the requirements of making superior quality holes in polymer composites. The vibration-assisted drilling has many exclusive features such as impacting, separating, changing speed and changing angle during the machining operation [39]. There are four different types of vibration-assisted drilling pro-cesses, namely (1) vibration-assisted twist drilling, (2) rotary ultrasonic machin-ing, (3) rotary ultrasonic elliptical machining and (4) conventional ultrasonic drilling. The former three drilling processes are *pulse-intermittent contact*-type machining process, that is, the tool comes intermittently in contact with the workpiece. In these processes, a high frequency and low amplitude vibration are superimposed on the tool along the feed direction [40]. The later one (conven-tional ultrasonic drilling) is a *non-contact*-type machining process, that is, there is no mechanical contact between the tool and workpiece. In ultrasonic drilling process, a high-frequency ultrasonic vibration is exposed on the tool along its longitudinal axis followed by the flow of abrasive slurry into the machining zone. The hammering and impact of the abrasive particles result in removal of material from the workpiece [41]. The ultrasonic machining can produce superior quality surface in the materials having hardness more than 40 HRC [42]. The schematic of the ultrasonic machining is shown in Figure 12.5. The complete experimental setup consists of power supply unit, mill-module assembly and abrasive slurry flow system. Table 12.1 shows some salient characteristics of the various modes of drilling processes.

In the past few years, vibration-assisted twist drilling has received extensive interest as this technique can reduce the thrust force by 20%–30% as compared to the conventional drilling [43]. The other works concluded that there is a significant

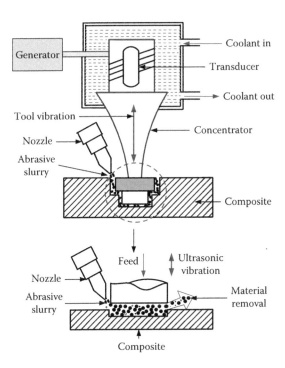

FIGURE 12.5   Schematic illustration of ultrasonic machining.

TABLE 12.1
Comparison of Various Modes of Drilling Processes

Aspects	CD	VATD	MAD	RUM	RUEM	CUD/RMUD
Interaction between tool and workpiece	Continuous	Pulse-intermittent				Non-contact
Material removal mechanism	Plastic deformation, shearing and bending rupture			Vibration-assisted diamond grinding process		Brittle fracture and plastic deformation due to hammering and impacting action of abrasive particles
Thrust force and torque	High	Thrust force and torque is reduced by 20%–30%		Low as compared to CD		Data is not available
Chip formation	Short and long chips	Short chips		Chip is removed majorly in the form of slug		Chips are removed in the form of micron size debris
Material removal rate		High		Low as compared to CD	9% lower as compared to CD	Low as compared to all the processes
Tool life	Low	High		High		Comparative data is not available
Surface roughness	High	High	Data not available	Low as compared to CD		
Quality of the drilled hole or damage produced during drilling	• Delamination • Splintering • Fibre pull-out • Matrix cracking • Debonding • Thermal damage etc	• Hole quality is improved • Delamination is reduced				• No delamination (thermosetting composites) • Minor fuzz at exit (thermoplastic composites) • No thermal, chemical, electrical or metallurgical threat to the workpiece

CD: Conventional Drilling, VATD: Vibration-Assisted Twist Drilling, MAD: Modulation-Assisted Drilling, RUM: Rotary Ultrasonic Machining, CUD: Conventional Ultrasonic Drilling, RUEM: Rotary Ultrasonic Elliptical Machining, RMUD: Rotary Mode Ultrasonic Drilling.

improvement in tool life during vibration-assisted twist drilling of composites as compared to the conventional drilling [44–48]. Takeyama and Kato [49] stated that burr-free holes can be effectively produced by means of ultrasonic vibration drilling using radial peripheral lip drill in glass fibre-reinforced composites. Takeyama and Iijima [50] found that the burr formation and surface damage can be prevented by superimposing ultrasonic vibration on the drill bit. Cong and co-workers [51] revealed that the thrust force, torque, surface roughness and delamination during rotary ultrasonic machining are lower than that of twist drilling. The authors have also reported that rotary ultrasonic machining can produce more than 1400 holes without wear of the tool, whereas twist drill bit can produce only 5 holes as the wear of the twist drill bit is rapid. Garlasco [52] reported that the life of the tool is improved by 10%–20% during rotary ultrasonic machining of boron fibre-reinforced epoxy composites as compared to the conventional drilling. Liu et al. [53] investigated the rotary ultrasonic elliptical machining of carbon fibre-reinforced composite panels using diamond core drill bits. It was reported that the defects encountered during drilling with this method result in significantly low damage to the hole as compared to the conventional drilling. Hocheng and Hsu [54] revealed that ultrasonic drilling can produce damage-free holes in comparison with the conventional drilling of polymer composites. It was also reported that the holes produced during ultrasonic drilling of carbon fibre-reinforced PEEK composites result in the production of high-quality holes as compared to the ultrasonic drilling of carbon fibre-reinforced epoxy composites. Brittle fracture of fibres and plastic deformation of the matrix are the two modes of material removal during ultrasonic drilling of composites. The slurry-assisted rotary mode ultrasonic drilling of glass fibre-reinforced epoxy composites using the hollow tool has been experimentally investigated [55]. The unique characteristics of rotary mode ultrasonic drilling method are shown in Figure 12.6. This newly developed method is a *contact-free* machining method, which is able to produce clean-cut holes without delamination. Moreover, the material removal rate was significantly increased during rotary mode ultrasonic drilling as compared to the conventional ultrasonic drilling method. The material removal

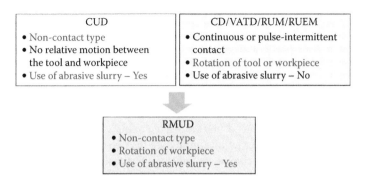

FIGURE 12.6 Characteristics of various modes of drilling processes. CD, conventional drilling; CUD, conventional ultrasonic drilling; VATD, vibration-assisted twist drilling; RUM, rotary ultrasonic machining; RUEM, rotary ultrasonic elliptical machining; RMUD: rotary mode ultrasonic drilling.

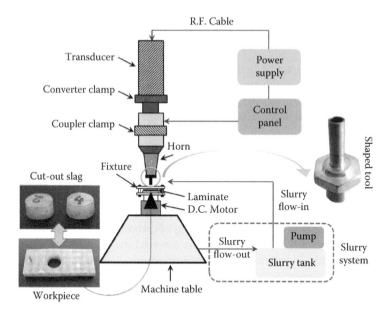

FIGURE 12.7 Schematic representation of rotary mode ultrasonic drilling and its components (R.F. Cable: Radio Frequencies Cable and D.C. Motor: Direct Current Motor).

rate was found to be increased linearly with the power rating, slurry concentration and abrasive particle size. Figure 12.7 represents the schematic of the rotary mode ultrasonic drilling.

12.7 CHALLENGES IN ADVANCED MACHINING OF POLYMER COMPOSITES

Although advanced machining techniques can be utilised as a viable alternative to the conventional machining process, there are some limitations posed by these techniques in the context of machining of polymer composites. The advanced machining is not suitable for all types of polymer composites. For instance, the electric discharge machining is particularly suitable for electrically conductive polymer composites such as carbon fibre-reinforced composites. However, the metallic filler may be incorporated to increase the electrical conductivity of the polymer composites to make it suitable for electric discharge machining. The electrochemical machining is a better choice for simultaneous drilling of multiple holes of different shapes in comparison to other advanced machining techniques. On the other hand, laser beam machining and the water jet machining are the two extensively used advanced machining techniques in the composite industry. The laser beam machining is suitable for aramid fibre-reinforced composites, resulting superior surface integrity. On the contrary, carbon fibre-reinforced composites are least suitable for laser beam machining. The major problem associated with laser beam machining is the thermal damage to the machined component because of high-heat generation during the machining processes, whereas the major difficulty associated with the water jet machining is the delamination that occurs due to the impact of the high-pressure water jet. However, the delamination can be reduced by optimising the setting of input parameters. The vibration-assisted machining results in improved tool life but the additional machining cost is much higher, whereas ultrasonic machining is particularly useful for hard and brittle polymer composites such as glass and carbon fibre-reinforced composites.

12.8 SUMMARY

The machining behaviour of polymer composites is substantially different from monolithic materials due to their anisotropic and inhomogeneous nature. The major problems associated with the conventional machining of polymer composites are severe damage to the machined surface and excessive wear of the tool. These are mainly encountered due to the generation of cutting forces during the machining. Therefore, the advanced machining techniques have been adopted by the composite fraternity as these processes offer an opportunity to process polymer composites economically, thus realising the full potential of the polymer composites. The notable features of the advanced machining techniques have been the absence of residual stresses, no heat generation and excellent surface finish, which make these techniques more attractive for machining of polymer composites. The advanced machining techniques are the potential and suitable solution for machining of polymer composites in many aspects as compared to the conventional machining techniques. But the development of cost-effective and damage-free machining method is yet a challenge for scientist and researchers.

REFERENCES

1. Debnath, K. and Singh, I. 2017. Low-frequency modulation-assisted drilling of carbon-epoxy composite laminate. *J. Manuf. Process.* 25:262–273.
2. Komanduri, R. 1997. Machining of fiber-reinforced composites. *Mach. Sci. Technol.* 1(1):113–152.
3. Ho, K.H. and Newman, S.T. 2003. State of the art electrical discharge machining (EDM). *Int. J. Mach. Tools Manuf.* 43:1287–1300.
4. Nakamura, M., Kanavama, K., and Hirai, Y. 1989. Electro-discharge machining of transformation toughened ZrO-NbC ceramic composite. *Mater. Manuf. Process.* 4(3):425–437.
5. Konig, W., Dauw, D.F., Levy, G., and Panten, U. 1988. EDM-Future steps towards the machining of ceramics. *CIRP Ann. Manuf. Techn.* 37(2):623–631.
6. Lau, W.S., Wang, M., and Lee, W.B. 1990. Electrical discharge machining of carbon fibre composite materials. *Int. J. Mach. Tools Manuf.* 30(2):297–308.
7. Guu, Y.H., Hocheng, H., Tai, N.H., and Liu, S.Y. 2001. Effect of electrical discharge machining on the characteristics of carbon fiber reinforced carbon composites. *J. Mater. Sci.* 36(8):2037–2043.
8. George, P.M., Raghunath, B.K., Manocha, L.M., and Warrier, A.M. 2004. EDM machining of carbon-carbon composite—a Taguchi approach. *J. Mater. Process. Technol.* 145(1):66–71.
9. Tandon, S., Jain, V.K., Kumar, P., and Rajurkar, K.P. 1990. Investigations into machining of composites. *Precis. Eng.* 12(4):227–238.

10. Singh, I. and Debnath, K. 2015. Advanced machining techniques for fiber-reinforced polymer composites. In *Processing Techniques and Tribological Behavior of Composite Materials*, R. Tyagi and J.P. Davim, Eds., pp. 317–340. Hershey, PA: IGI Global.

11. Abrate, S. and Walton, D. 1992. Machining of composite materials. Part II: Non-traditional methods. *Compos. Manuf.* 3(2):85–94.

12. Cook, N.H., Foot, G.B., Jordan, P., and Kalyani, B.N. 1973. Experimental studies in electro machining. *J Eng. Ind.* 94(4):945–950.

13. Jain, V.K., Sreenivasa Rao, P., Choudhary, S.K., and Rajurkar, K.P. 1991. Experimental investigations into travelling wire electro-chemical spark machining (TW-ECSM) of composites. *J. Eng. Ind.* 113(1):75–84.

14. Jain, V.K., Tandon, S., and Kumar, P. 1990. Experimental investigations into electro-chemical spark machining of composites. *J. Eng. Ind.* 112(2):194–197.

15. Adams, R.B. 1986. Water jet machining of composites. SME Paper EM 86-113.

16. Howarth, S.G. and Strong, A.B. 1990. Edge effects with water jet and laser beam cutting of advanced composite materials. In *Proceedings of the 35th International SAMPE Symposium*, CA, United States: Society for the Advancement of Material and Process Engineering, pp. 1685–1697.

17. Sprow, E.E. 1987. Cutting composites: Three choices for any budget. *Tool. Prod.* 43(12):46–50.

18. Ramulu, M. and Arola, D. 1993. Water jet and abrasive water jet cutting of unidirectional granite/epoxy composite. *Composites* 24(4):299–308.

19. Hashish, M. 1989. Machining of advanced composites with abrasive waterjets. *Manuf. Rev.* 2(2):142–150.

20. Hashish, M. 1984. A modelling study of metal cutting with abrasive waterjets. *J. Eng. Mater. Technol.* 106(3):88–100.

21. Kim, T.J., Sylvia J.G., and Posner, L. 1985. Piercing and cutting of ceramics with abrasive-waterjets. *PED-vol. 17*, New York: ASME, pp. 19–24.

22. Shanmugam, D.K., Nguyen, T., and Wang, J. 2008. A study of delamination on graphite/epoxy composites in abrasive waterjet machining. *Compos. A* 39(6):923–929.

23. Ho-Cheng, H. 1990. Failure analysis of water jet drilling in composite laminates. *Int. J. Mach. Tools Manuf.* 30(3):423–429.

24. Hamatani, G. and Ramulu, M. 1990. Machinability of high temperature composites by abrasive waterjet. *J. Eng. Mater. Technol.* 112(4):381–386.

25. Hashish, M. 1993. Precision drilling of composites with abrasive waterjets. *ASME Bound Volume*, vol. 45, pp. 217–225.

26. Azmir, M.A. and Ahsan, A.K. 2008. Investigation on glass/epoxy composite surfaces machined by abrasive water jet machining. *J. Mater. Process. Technol.* 198(1):122–128.

27. Azmir, M.A. and Ahsan, A.K. 2009. A study of abrasive water jet machining process on glass/epoxy composite laminate. *J. Mater. Process. Technol.* 209(20):6168–6173.

28. Flaum, M. and Karlsson, T. 1987. Cutting fiber-reinforced polymers with CW CO_2 laser. In *SPIE-High Power Lasers and Their Industrial Applications*, vol. 801, pp. 142–149.

29. Powell, J., Ellis, G., Young, C.D., Menzies, I.A., and Scheyvaerts P.F. 1987. CO_2 laser cutting of non-metallic materials. In *Proceedings of 4th International Conference on Lasers in Manufacturing*, Birmingham, UK, pp. 69–82.

30. Chryssolouris, G., Bredt, J., and Kordas, S. 1985. Laser turning for difficult to machine materials. *ASME* PED-vol. 17:9–17.

31. Vanderwert, T.L. 1983. Machining plastics with lasers. *Manuf. Eng.* 91(5):55–58.

32. Lawson, W.E. 1986. Laser cutting of composites. *Composites in Manufacturing*, Los Angeles, CA: Society of Manufacturing Engineers, pp. 1–10.

33. Lau, W.S. and Lee, W.B. 1991. A comparison between EDM wire-cut and laser cutting of carbon fiber composite materials. *Mater. Manuf. Process.* 6(2):331–342.

34. Mathew, J., Goswami, G.L., Ramakrishnan, N., and Naik, N.K. 1999. Parametric studies on pulsed Nd:YAG laser cutting of carbon fiber reinforced plastic composites. *J. Mater. Process. Technol.* 89:198–203.
35. Tagliaferri, V., Di-Rio, A., and Crivelli Visconti, I. 1985. Laser cutting of fibre-reinforced polyesters. *Composites* 16(4):317–325.
36. Mello, M.D. 1986. Laser cutting of non-metallic composites. In *Proceedings of SPIE - Laser Processing: Fundamentals, Applications, and Systems Engineering*, vol. 668, Canada: The International Society for Optical Engineering, pp. 288–290.
37. Young, T.M. 2008. Impact of Nd-YAG laser drilling on the fatigue characteristics of APC-2A/AS4 thermoplastic composite material. *J. Thermoplast. Compos. Mater.* 21(6):543–555.
38. Yung, K.C., Mei, S.M., and Yue, T.M. 2002. A study of the heat-affected zone in the UV YAG laser drilling of GFRP materials. *J. Mater. Process. Technol.* 122(2):278–285.
39. Wang, L.J. and Ji, Z. 1987. Kinematics and surface roughness in turning with ultrasonic vibration tool. *J. Arms.* 9(8):24–31.
40. Liu, D., Tang, Y., and Cong, W.L. 2012. A review of mechanical drilling for composite laminates. *Compos. Struct.* 94(4):1265–1279.
41. Nath, C., Lim, G.C., and Zheng, H.Y. 2012. Influence of the material removal mechanisms on hole integrity in ultrasonic machining of structural ceramics. *Ultrasonics* 52(5):605–613.
42. Singh, R. and Khamba, J.S. 2006. Ultrasonic machining of titanium and its alloys: A review. *J. Mater. Process. Technol.* 173(2):125–135.
43. Zhang, L.B., Wang, L.J., and Wang, X. 2003. Study on vibration drilling of fiber reinforced plastics with hybrid variation parameters method. *Compos. B* 34:237–244.
44. Ramkumar, J., Aravindan, S., Malhotra, S.K., and Krishnamurthy, R. 2004. An enhancement of the machining performance of GFRP by oscillatory assisted drilling. *Int. J. Adv. Manuf. Technol.* 23(3–4):240–244.
45. Ramkumar, J., Malhotra, S.K., and Krishnamurthy, R. 2004. Effect of workpiece vibration on drilling of GFRP laminates. *J. Mater. Process. Technol.* 152(3):329–332.
46. Arul, S., Vijayaraghavan, L., Malhotra, S.K., and Krishnamurthy, R. 2006. The effect of vibratory drilling on hole quality in polymeric composites. *Int. J. Mach. Tool. Manuf.* 46(3):252–259.
47. Wang, X., Wang, L.J., and Tao, J.P. 2004. Investigation on thrust in vibration drilling of fiber reinforced plastics. *J. Mater. Process. Technol.* 148(2):239–244.
48. Zhang, L.B., Wang, L.J., Liu, X.Y., Zhao, H.W., Wang, X., and Luo, H.Y. 2001. Mechanical model for predicting thrust and torque in vibration drilling fiber-reinforced composite materials. *Int. J. Mach. Tool. Manuf.* 41(5):641–657.
49. Takeyama, H. and Kato, S. 1991. Burrless drilling by means of ultrasonic vibration. *CIRP Ann.* 40(1):83–86.
50. Takeyama, H. and Iijima, N. 1988. Machinability of glass fiber reinforced plastics and application of ultrasonic machining. *CIRP Ann.* 37(1):93–96.
51. Cong, W.L., Pei, Z.J., Feng, Q., and Deines, T.W. 2012. Rotary ultrasonic machining of CFRP: a comparison with twist drilling. *J. Reinf. Plast. Compos.* 31:313–321.
52. Garlasco, F.M. 1969. Machining of boron fibrous composites. *SME Technical Paper EM 69-14*, United States: American Society of Mechanical Engineers.
53. Liu, J., Zhanga, D., Qin, L., and Yan, L. 2012. Feasibility study of the rotary ultrasonic elliptical machining of carbon fiber reinforced plastics (CFRP). *Int. J. Mach. Tool. Manuf.* 53(1):141–150.
54. Hocheng, H. and Hsu, C.C. 1995. Preliminary study of ultrasonic drilling of fiber reinforced plastics. *J. Mater. Proc. Technol.* 48(1–4):255–266.
55. Debnath, K., Singh, I., and Dvivedi, A. 2015. Rotary mode ultrasonic drilling of glass fiber-reinforced epoxy laminates. *J. Compos. Mater.* 49(8):949–963.

13 Joining Behaviour of Fibre-Reinforced Polymer Matrix Composites

Inderdeep Singh, J. Kumar and U.K. Komal

CONTENTS

13.1   INTRODUCTION

The use of polymer matrix composites is nowadays increasing in many applications including automotive, aerospace and household products. Light weight and high strength-to-weight ratio enhance their demand in structural and non-structural applications. There are several techniques available to develop composite parts such as hand lay-up, compression moulding and injection moulding. These processes require optimisation of parameters for fabricating different types of polymer matrix composites.

On the basis of occurrence, fibres are divided into two categories: (1) synthetic fibre and (2) natural fibre. Synthetic fibres are obtained from petroleum-based resources, such as glass and carbon fibre. Synthetic fibre-reinforced composites are extensively used in many engineering applications, but these materials face problems because petroleum-based resources are depleting rapidly. Another problem with synthetic fibre is its non-biodegradability. Due to this, researchers are focusing towards the use of natural fibres as reinforcement. Natural fibres are biodegradable in nature, and resources of natural fibre are not limited like synthetic fibre. Recently, natural fibre-reinforced polymer matrix composites have been used for non-structural applications in automotive and aerospace industries. Major source of natural fibres is plant. Some common plant fibres that are being used in the industry include jute, sisal, nettle, hemp, kenaf, flax and bamboo.

Thermosets and thermoplastics, two different kinds of polymers, are used as matrix in polymer composites. Both are obtained from petroleum-based resources. But in future, due to limited stock of petroleum resources, these materials are going to be extinct. Therefore, researchers are focusing on development of biopolymers. These are obtained from synthesis of natural resources. Polylactic acid (PLA), polyhydroxy-alkanoate (PHA) and poly(3-hydroxybutyrate-co-3-hydroxyvalerate) (PHBV) are biopolymers, which are thermoplastics. These polymers are biodegradable in nature. Thermosets carry cross linking polymeric chain, are irreversible in nature (after curing cannot return to its initial state), less elastic, chemically resistant and non-recyclable. On the other hand, thermoplastics carry linear chain, are more elastic, reversible in nature and recyclable.

Joining of polymer matrix composites seems to be simple, but it is a very challenging task. Improper joint design leads to stress concentration in joint section and may become cause of failure during service life of the product. Therefore, study of joining behaviour of polymer matrix composites is necessary. There are mainly three basic methods available for joining of polymer matrix composites: adhesive joining, mechanical fastening and fusion bonding. Selection of an individual method depends on the application of the final product, where it is going to be used. For non-structural applications, adhesive and fusion joining are preferable because load acting on joint section is less. For structural applications, mechanical fastening is preferable because of its high load-carrying capacity.

Many researchers published articles on joining behaviour of synthetic fibre-reinforced polymer composites by using all the three joining techniques. Some researchers have studied joining behaviour of hybrid composites, such as composite to metal and composite to plastic. Advance joining techniques such as friction stir welding, induction joining, microwave joining have also been reported in the literature for synthetic fibre-reinforced composites. But articles focusing on the joining behaviour of natural fibre-reinforced composites are not much.

13.2   NEED OF JOINING

Most of the polymer composite parts are fabricated to a near-net shape, but in order to obtain an intricate structure, several composite components have to be joined together. Therefore, individual components are formed separately and then assembled. Joining process requires various operations (machining, surface preparation, heating) to form a joint, which causes stress concentration at the joint interface. Therefore, maximum chance of failure of material occurs at the joint interface.

The requirements for joining are as follows:

1. To fabricate large and intricate composite products by assembling individual composite parts.
2. To provide maintenance, repair and service to product. A product having only one part is difficult to maintain and service. After any damage or failure, complete product has to be replaced. In case of assembled product, any defected component can be easily removed and repaired.

13.3   ADHESIVE JOINING

Adhesive joining is a joining process in which two similar and dissimilar parts are joined together by means of an adhesive. Part that is going to be joined is called as adherend. Basically two types of adhesives are used: thermoset and thermoplastic. For joining thermoset matrix composites, thermoset adhesives are compatible; for thermoplastic matrix composites, thermoplastic adhesives are used. Thermosets have more number of polar groups (unsaturated bond) in their polymeric chain. During surface preparation, these polar groups are exposed to surface, which helps thermoset to make proper bonding with adhesives (Baldan 2004). On the other hand, thermoplastics have long and saturated polymeric chain structure. These polymeric chains have less number of polar groups as compared to thermosets. It causes weak bonding between thermoplastics and adhesives. To enhance the surface property of thermoplastics, advance surface preparation techniques are required such as plasma treatment, corona discharge treatment and ion treatment (Davies et al. 1991). The treatment increases the free radicals on the surface of the thermoplastic, which improves the bonding property of thermoplastic with adhesive. Another method to enhance the bonding of thermoplastic with adhesive is use of solvent at the joining surface, which chemically reacts with surface and softens the thermoplastic, and finally pressure is applied between the joining interfaces, which lead to the formation of strong bond between adherend and adhesive.

Steps involved in joining of polymer matrix composites:

1. Fabrication of composite parts
2. Selection of adhesive
3. Selection of joint design
4. Surface preparation of adherend
5. Spreading of desired amount of adhesive on selected design configuration
6. Application of pressure and temperature for proper consolidation
7. Curing of joint

Joint design depends on type of application and loading condition of final product. There are different types of joint configurations used in adhesive joining. Lap joint, butt joint, strap joint, step joint and scarf joint are examples of common types of adhesively bonded joint configurations. Figure 13.1 shows all different types of joint configurations. Single-lap joint is most commonly used joint design because of its simple design configuration. Off-centring of load acting on the adherend is the main drawback, which causes the cleavage stress that further lead to the failure of the joint. To prevent this type of problem, double-lap joint configuration is used. For more efficient joint, step and scarf joints are preferred.

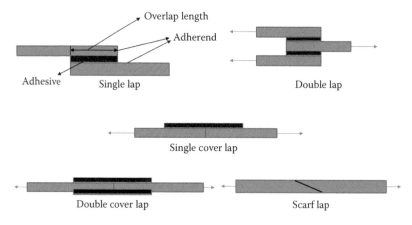

FIGURE 13.1 Different types of adhesive joints.

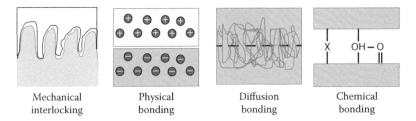

| Mechanical interlocking | Physical bonding | Diffusion bonding | Chemical bonding |

FIGURE 13.2 Bonding mechanism in adhesive joining.

13.4   BONDING MECHANISM IN ADHESIVE JOINING

There are four basic mechanisms: mechanical interlocking, physical bonding, diffusion and chemical bonding by which adhesive bonding takes place (Messler 2004). The schematic representation of the mechanism is shown in Figure 13.2.

All the four mechanisms or combination of one or two mechanisms may participate during the adhesive joining process. Bonding mechanism mainly depends on material properties such as surface energy, surface texture and reactive nature of adhesive and adherend.

13.4.1   MECHANICAL INTERLOCKING

Adherend surface consists of irregularities and voids at microscopic level. These irregularities and voids are responsible for mechanical interlocking. When adhesive is applied on adherend surfaces, it wets the surface and fills voids and irregularities present on the surface. Further, application of pressure removes air entrapped in cavities so that the adhesive fills the cavities properly. After curing, adhesive forms interlocking with adherend surface, which provides joint strength. Therefore, joint strength depends on irregularities and cavities present on adherend surface. This can be increased by increasing the surface roughness of adherend surface. Mechanical abrasion, sanding and grit blasting are the processes to increase the surface roughness of adherend surface. In mechanical abrasion and sanding, different grades of emery paper are used to roughen the surface; in grit blasting, abrasives are used to roughen the adherend surface.

13.4.2   PHYSICAL BONDING

Physical bonding between the adherend and adhesive takes place in presence of Van der Waal forces, hydrogen bonding, dipolar interactions and other low-energy forces at molecular level. It happens due to the difference in electronegativity characteristics between adherend and adhesive.

13.4.3   DIFFUSION BONDING

Partial solubility of material is the necessary condition for diffusion bonding. Therefore, it is mainly applicable for thermoplastics. They can only transform into partial and fully soluble phase; soluble phase is obtained by using chemical or solvent at joining interface. Diffusion bonding efficiency is better when similar types of adhesive and adherend materials are used.

13.4.4 CHEMICAL BONDING

Chemical force is the primary force in adhesive joining; this leads to chemical bonding between adherend and adhesive at the joining interface. Bonding strength depends on surface energy of adherend material; high surface energy on adherend surface gives higher bonding strength and good wettability.

13.5 TYPES OF ADHESIVES

Various categories of adhesives are as follows:

- Natural and synthetic adhesives
- Organic and inorganic adhesives
- Structural and non-structural adhesives
- Classification on the basis of chemical composition
- Classification on the basis of physical form
- Classification on the basis of mode of application, curing and setting mechanism

On the basis of occurrence, adhesives are classified into natural and synthetic adhesives. Natural adhesives are obtained from plant and animal sources. Some common examples are pine tar, starch, casein, albumin and shellac. Synthetic adhesives are chemically synthesised.

Organic adhesives are based on polymer, and inorganic adhesives are based on metallic and non-metallic compounds. In terms of strength and thermal stability, inorganic adhesives are more stable than organic adhesives. Use of structural and non-structural adhesives is based on strength and durability. For holding heavy-loaded structure, structural adhesives are used; for vibration damping, impact absorption, thermal and electrical insulation, non-structural adhesives are used. Adhesives can also be classified by their chemical structure; according to this, adhesives are divided into three categories: (1) thermoset adhesive, (2) thermoplastic adhesive and (3) elastomeric adhesive. Adhesives are available in different forms such as paste, liquid, powder, tape and film. Therefore, they may be classified based on their physical states. Mode of application of adhesive on adherend surface can be done in different manners. It can be sprayable, brushable, trowelable or extrudable. Curing and setting mechanism of different adhesives are different. Some need heat; some need pressure; few are cured at room temperature and others may need catalyst to cure or set.

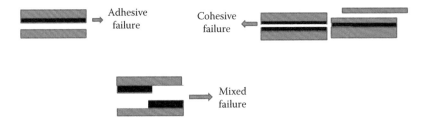

FIGURE 13.3 Failure mechanism of adhesive joint.

13.6 FAILURE MECHANISMS

Adhesive joints can fail by three mechanisms: adhesive failure, cohesive failure and mixed (adhesive and cohesive) failure. All three modes of failure are shown in Figure 13.3. Improper surface preparation, wrong adhesive selection and high peel stress cause adhesive failure. It happens at the adhesive–adherend joint interface. In cohesive mode of failure, either adherend fails or adhesive fails. It is due to high bond strength between adhesive and adherend than their individual strength. In mixed type of failure, adhesive and cohesive failures take place simultaneously.

13.7 ADVANTAGES OF ADHESIVE JOINING

- In adhesive joining, load is uniformly distributed at joint section except at the corner (lap joint).
- Stress concentration in adhesive joining is lesser than mechanical joining. Therefore, it can take more flexural and fatigue load as compared to mechanical joints.
- Adhesive joints can absorb damp vibration and shock waves.
- Adhesive joints provide sealing.
- They can be used to join thin or thick adherends.
- They act as insulators (heat and electricity).
- Large surface can be easily joined.

13.8 DISADVANTAGES OF ADHESIVE BONDING

- Permanent joint
- Stress analysis is difficult
- Replacement of damaged parts is not possible
- Improper surface preparation causes joint failure
- Inspection of joints is difficult
- Long curing time needs long time to set the adhesive
- Adhesives are sensitive to some chemical solvent and the environment
- Sensitive to high-temperature applications

13.9   MECHANICAL JOINING

Mechanical joining is the oldest method used for joining of metallic parts. It can also be used for joining of composites. It ensures interlocking or interference force between mating surfaces of joined components to prevent undesirable movement and unintentional disassembly. When an extra supplement (mechanical fastener) is required for interference forces and subsequent interlocking, then it is called mechanical fastening. When geometric features are provided to mating surface for interference forces and interlocking, then it is called integral mechanical attachment. Further mechanical fastening is divided into two types on the basis of fasteners being used: (1) threaded fastener and (2) unthreaded fastener. Threaded fasteners require preloading to get interference between mating parts, and unthreaded fasteners provide interference by bearing action. For the application of mechanical fastener, hole is required in the joining specimen. Drilling a hole in composite is a challenging task because composites are usually heterogeneous and anisotropic. It depends on types of fibre, types of matrix, stacking sequence of laminates and processing techniques. Therefore, selection of a best drilling parameter without any initial optimisation is difficult.

Drilling causes several defects in fibre-reinforced composites, such as spalling, fuzzing, delamination and edge chipping that can be visualised with naked eye (Konig et al. 1985). Defects such as crack formation and damage on peripheral zone are visualised under microscope. Drilling-induced defects depend on types of fibre, tool geometry, drilling parameters and drilling environment. Most of the defects occur due to the effect of tool geometry during drilling operation because fibres such as carbon and glass (mostly used in composite as reinforcement) are abrasive in nature. At the time of drilling, when tool makes contact with fibres, it gets eroded and its dimensions change. An eroded tool generates more drilling forces while drilling a hole, which apply compressive force rather than shear force that results in damaged and defective hole. But the major benefit of mechanically fastened joint is that it can be disassembled, repaired and easily inspected.

13.9.1   ADVANTAGES OF MECHANICAL JOINING

* There is no need of surface preparation.
* The joints provide maintenance, repair, inspection and service.
* It gives easy disassembly and assembly of joints.
* It can be used for metallic, non-metallic and combination of metallic and non-metallic materials.
* Property of parent material remains unchanged after joining.
* Thermal stress is absent at joint interface.

13.9.2   DISADVANTAGES OF MECHANICAL JOINING

* Hole drilling causes stress concentration at joint section.
* The fastener increases the weight of structure.

- Near drilled section, fibres are open to atmosphere. If fibres are hydrophilic, they absorb moisture from atmosphere and cause swelling of joint section.
- Common types of defects observed in composite parts with hole include spalling, fuzzing and delamination.

13.10   FUSION BONDING

Fusion bonding is a joining process, mainly applicable for thermoplastics. It is not applicable for thermosets because they cure irreversibly. Once thermoset is cured, it cannot return to its original state, but thermoplastics are reversible in nature. Therefore, thermoplastic-based composites can be easily joined by fusion-bonding technique. In thermoplastics, polymeric molecules form long and straight chain. These polymeric chains are held together by secondary bonds (weak bond) as shown in Figure 13.4.

When the thermoplastic is heated, the secondary bonds break, and the polymer chains get mobile; mobility of chain increases with an increase in the temperature. These polymeric chains take part in bond formation at joining interface. Thermoplastics are divided into two categories: amorphous plastics and semi-crystalline plastics. In amorphous plastics, polymeric chains are arranged in random manner as shown in Figure 13.5.

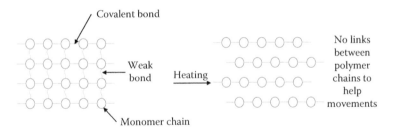

FIGURE 13.4   Polymeric chains in thermoplastic before and after heating.

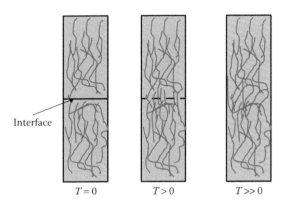

FIGURE 13.5   Bonding in amorphous thermoplastics.

When heat is supplied, polymeric chain in amorphous region gets mobile at glass-transition temperature of thermoplastic and crosses the joint interface. Further, polymeric chains of one adherend entangle or diffuse with polymeric chain of other adherend.

Semi-crystalline thermoplastics contain both crystalline and amorphous region. Here, polymeric chains are arranged in a particular manner in crystalline region, and chains are randomly distributed in amorphous region. When heat is supplied, polymeric chain in amorphous region gets mobile at glass-transition temperature, but crystalline phase needs more amount of heat for mobility of arranged polymeric chain. At the melting point, arranged polymeric chains in crystalline region loss their strength and start moving. Figure 13.6 shows mechanism of bonding in semi-crystalline thermoplastic.

In fusion bonding, heat can be generated at the interface of joining specimen by thermal, friction and electromagnetic means, which are mentioned in Figure 13.7.

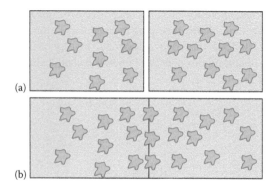

FIGURE 13.6 Bonding in semi-crystalline thermoplastics: (a) Before heating and (b) after heating.

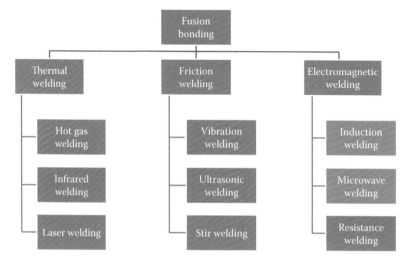

FIGURE 13.7 Classification of fusion-bonding techniques.

Infrared, laser and hot gas are used in thermal welding. Vibration, ultrasonic and stirring mechanism are used in friction welding. Similarly, induction, microwave and resistance heating are used in electromagnetic welding technique. Among all the three modes of heating, electromagnetic heating is fast and can be controlled.

13.10.1 Bonding Mechanism during Fusion Bonding

Bonding between thermoplastic polymers at joining interface takes place by three modes: (1) chemical bonding, (2) chain entanglement and (3) diffusion of chain (Awaja 2016). The schematic representation of the mechanics is shown in Figure 13.8.

When two polymer surfaces of joining specimen are brought into intimate contact and heated above their glass-transition temperature, chain entanglement, diffusion or chemical bonding take place at joining interface. Entanglement of polymeric chain depends on molecular weight and physical properties of interface of the thermoplastic, and interdiffusion depends on crystallinity, viscosity and molecular weight of the thermoplastic. Chemical bonding takes place due to bond formation between polymeric chains of joining interfaces.

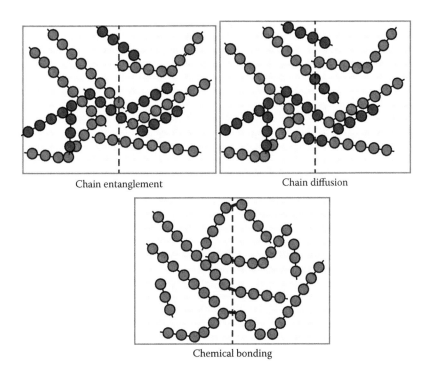

Chain entanglement   Chain diffusion

Chemical bonding

FIGURE 13.8 Bonding mechanism in fusion-bonding process.

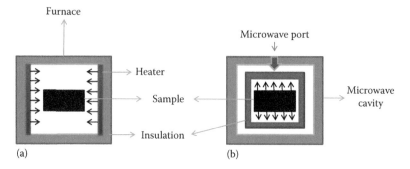

FIGURE 13.9   (a) Conventional versus (b) microwave heating mechanism.

13.10.2 MICROWAVE JOINING

Microwave joining is a fusion-bonding technique where the material is heated by microwaves. Microwaves consist of electromagnetic waves. Here, electric and magnetic fields are mutually perpendicular to each other. When material is exposed to microwaves, both fields interact with material and cause joule loss, hysteresis loss, dipolar loss, conduction loss, eddy current loss, dielectric loss or residual loss. These losses depend on magnetic and electrical properties of material, and it takes place at molecular level. These different types of losses come out from the material in the form of heat. Due to this reason, microwave heating is different from conventional heating. In conventional heating, heat is supplied to material from external source, so heat flows in material from outside to inside but in microwave heating, heat flows in material from inside to outside. It is shown in Figure 13.9.

13.10.3 HEATING MECHANISM IN MICROWAVE HEATING PROCESS

Heating mechanism depends on magnetic and electrical properties of the materials. There are four kinds of materials in terms of magnetic and electrical conductivity: (1) non-magnetic and non-conductive, (2) non-magnetic and conductive, (3) magnetic and non-conductive and (4) magnetic and conductive. All the four kinds of materials behave differently when exposed to microwave radiation. Therefore, heating mechanisms are also different.

13.10.3.1 Heating Mechanism in Non-Conductive and Non-Magnetic Material

Some materials such as pure water, ceramics, ceramic matrix composites, polymers, polymer matrix composites and food products are neither conductive nor magnetic in nature but in presence of electric field, dipoles are generated inside the material. If electric field is oscillating in nature, then dipoles re-orient themselves after each half oscillation to be in phase with oscillating electric field. This re-orientation causes friction between molecules and generates heat.

13.10.3.2   Heating Mechanism in Non-Magnetic and Conductive Material

Materials such as aluminium and copper are conductive and non-magnetic in nature. Here, heating of material takes place by conduction loss due to presence of free electrons in conductive material. When electric field is applied on these materials, electrons start to move in opposite direction of applied electric field, which causes generation of electric field inside the material. The direction of induced electric field is in opposite direction to the applied electric field. If the direction of the applied electric field changes, then induced electric field direction also changes. Therefore, shifting of direction of induced electric field causes friction between molecules and gives generation of heat inside the material.

13.10.3.3   Heating Mechanism in Magnetic and Non-Conductive Material

In non-conductive and magnetic material, heating of material takes place by the hysteresis loss and residual loss. Hysteresis loss is due to the disturbance induced by external magnetic field in oriented magnetic domain. Initially, magnetic domains are arranged in such a manner that overall magnetic effect of material is zero. When magnetic field is applied, domains align themselves along the direction of magnetic field. After releasing magnetic field, domains do not come to its initial position. This misalignment is a part of magnetic energy, which is further converted into heat. Residual loss is due to the displacement of domain-wall resonance or disturbance in the electron spin in presence of external magnetic field. In domain-wall resonance, domain parallel to magnetic field expands and non-parallel domain contracts. Further, reversal of magnetic field direction causes expansion of domain parallel to magnetic field and contraction of rest of the domain. Therefore, continuous contraction and expansion cause heating of material. Disturbance in electron spin is generally found in ferromagnetic materials.

13.10.3.4   Heating Mechanism in Magnetic and Conductive Material

If material is conductive and magnetic in nature, then heat generation in material takes place by the combined effect of conduction and magnetism. It includes all the heat loss mechanisms of conduction and magnetism. In microwave joining of thermoplastic composites, only joining section of composite is exposed to microwaves, and rest of the portion is masked. Masking is required to prevent the heating of whole sample. The main focus is to heat the joint section only. Therefore, it is also called selective microwave heating. Microwave joining of thermoplastic composites depends upon dielectric properties of thermoplastic. Thermoplastics having high dielectric loss factor generate large amount of heat at joint interface. Further, the dielectric properties of thermoplastic depend upon its crystallinity. Polymers with a degree of crystallinity above 45% are essentially transparent to microwaves due to the restriction of dipoles. As the degree of crystallinity increases, the dielectric loss factor decreases. Although thermoplastic polymers generally have low loss factors at room temperature, the addition of conductive fillers or fibres can strongly influence the overall dielectric loss. Some common examples are charcoal powder, metal mesh, metal powder, metal foil and carbon fibre prepreg. Yarlagadda and Chai (1998) investigated joining of thermoplastics using focused microwave energy. The thermoplastic joining process was done by using two methods. In the first case, microwave energy was directly induced to the specimen joint interface, whereas in the second case, primers in the form of epoxy-based resin were applied to promote the joining of the materials by means of microwave energy. Experiments were performed on ultra-high molecular weight polyethylene (UHMW PE), polycarbonate (PC) and acrylonitrile butadiene styrene (ABS) polymers under room temperature conditions. Further, tensile tests were conducted on joined sample. Mishra and Sharma (2016) concluded that the effective processing of polymer matrix composites depends upon the dipole structure, frequency of microwave, specimen temperature and characteristic properties of reinforcement and matrix materials. Kathirgamanathan (1993) investigated microwave joining of thermoplastic using conductive polymer. The welding took place in 2–120 s with lap shear strength of 19.0 ± 2.0 N mm^{-2}. Martinelli et al. (1985) found that thermoset matrix, in initial liquid phase, efficiently absorbs microwave at room temperature. However as the matrix gets set, the increasing viscosity restricts changes in dipole orientation in oscillatory electric field. This reduces microwave absorption in thermosets due to higher rate of cross linking. Singh et al. (2013) reported feasibility study of microwave joining of green composite. Composites were joined using charcoal as a microwave coupling agent. Bajpai et al. (2012) investigated joining behaviour of four different types of natural fibre composites with adhesive and microwave energy and simulated the microwave heating process with COMSOL Multiphysics software.

13.11   INDUCTION JOINING

Induction joining is also used for joining thermoplastics and thermoplastic-based composites. Here materials are joined by means of induction heating. The setup consists of induction coil, radio-frequency power generator, susceptor (in case of non-conducting fibre) and cooling mechanism. The schematic is shown in Figure 13.10.

Power generator supplies current and voltage to induction coil. After that, induction coil generates electromagnetic field, which interacts with susceptor material or

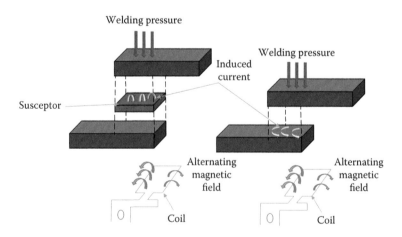

FIGURE 13.10   Induction heating with susceptor and without susceptor.

conducting fibre and generates heat at joining interface. If fibre is conducting in nature (carbon fibre), then there is no need of susceptor material; otherwise, susceptor is needed between joining interface to generate required amount of heat. Cooling system is also required to cool induction coil because it gets heated during the operation.

13.11.1 Heating Mechanism

The heating mechanism in induction joining process is simple. When an alternating voltage is supplied to conductive coil, an alternating current is generated. Further, this alternating current induces an alternating magnetic field with frequency similar to the alternating current. After that, alternating magnetic field interacts with susceptor, which is placed in the vicinity of coil. The susceptor material gets heated due to conduction loss, hysteresis loss or eddy current loss. This is all related to nonconductive reinforcement. For conductive fibres, heating mechanism is different. Mainly carbon fibres are available as conducting fibre, so most of the experiments and theory for heating mechanism are based on carbon fibre. Rudolf et al. (2000) gave induction heating mechanism based on joule losses in the carbon fibres when current is flowing in the laminate. The heat-affected area in the plane of the workpiece was found to be the mirror image of the coil. The coil orientation had no influence on the heating rate, and the time to heat the composite laminate to the desired temperature decreases quadratically with increasing frequency. Yarlagadda and Kim (2002) gave another theory for heating mechanism of carbon fibre-reinforced composites by induction. It was found that junction heating effects (dielectric hysteresis or contact resistance) dominate over fibre heating. It was also reported that conductive fibre architecture plays a critical role in determining the dominant mechanism. Woven fabrics may show fibre dominance due to direct fibre contact, whereas laminated prepreg-based systems may show dominance of junction heating. Fink et al. (1995) also reported another heating mechanism in induction welding for carbon fibre laminates. It was found that fibre–fibre contact is difficult in laminate as matrix is present between the fibres. So, dielectric heating plays an important role.

From earlier experimentation and theory, four types of heating mechanisms were proposed for induction heating: (1) joule loss/fibre heating, (2) junction heating/ dielectric hysteresis heating, (3) junction heating/contact resistance and (4) hysteresis/ eddy current loss. Figure 13.11 shows sites for different types of heating.

Joule loss/fibre heating (conducting fibre) is due to the resistance offered by fibre when current is flowing through it. Therefore, amount of heat generation is dependent upon the length of fibre, cross section of fibre and resistivity of fibre. Junction heating/dielectric hysteresis heating takes place in the matrix. At the joint interface, a thin layer of matrix is present between fibres, which forms dielectric medium. When matrix interacts with alternating electric field, dielectric loss takes place. This loss comes out in the form of heat. Junction heating/contact resistance occurs at fibre–fibre contact. It is found mainly in angle ply-type composite laminate where voltage drop takes place at fibre–fibre contact. Hysteresis/eddy current loss takes place if susceptor is used at joining interface, and susceptor is magnetic/ conductive in nature.

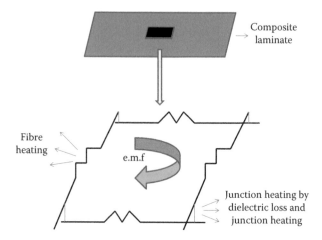

FIGURE 13.11   Induction heating mechanism having conducting fibre in reinforcement.

13.12   RESISTANCE WELDING

Resistance welding is a fusion-bonding technique, which is mainly used for thermo-plastic and thermoplastic matrix composites. Here, an electrically resistive implant is placed between the joining interfaces. When current is flowing through implant, it causes joule loss, which comes in the form of heat (Warren et al. 2016). Further, the heat generated at joining interface melts the thermoplastic. Thereafter, required amount of pressure is applied on specimen for proper consolidation. In this way, joining takes place between various types of thermoplastic and thermoplastic matrix composites. Resistance welding setup is shown in Figure 13.12.

It mainly consists of power supply, clamping device, implant, wires, voltmeter and ammeter. Thermocouples are also used in case of temperature measurement.

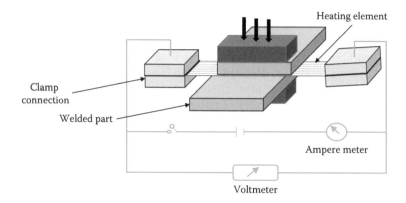

FIGURE 13.12   Resistance heating for thermoplastic composites.

Initially, all the electrical connections are done and then current is supplied. Current passes through implant and generates heat. Heat generation raises temperature at the joining interface. This should be near glass-transition temperature for amorphous plastics and melting temperature for semi-crystalline plastics. The heat generation should be in such a manner that heat-affected zone (HAZ) is small in specimen. Thereafter, current is switched off, and pressure is applied. Application of pressure removes gap, cavities and air bubbles at the joining interface. Further, it provides proper contact between joining interfaces. To prevent the heat loss from heating element, insulation of heating element is required. Insulation can be done by using different types of materials such as ceramic, asbestos, silicon rubber and teflon.

The heating element is a key factor in joining of thermoplastic and thermoplastic matrix composites by resistance welding. It is also called implant. It gives the necessary heat energy to the joint interface and is trapped inside the joint after welding. Any electrically conductive material can be used as implant. Generally, two types of materials are used as implants: one is carbon fibre, and another is stainless steel metal mesh.

13.13   ADVANTAGES OF FUSION WELDING

*   Joining process is fast; it takes very less time to join the specimen.
*   There is no need of surface preparation and any treatment of surface before joining.
*   Wide sources are available for joining.
*   Process can be automated.
*   It can be used at industrial scale.

13.14   DISADVANTAGES OF FUSION WELDING

*   Limited to thermoplastic-based composites.
*   Control on HAZ is difficult during joining process.
*   It can be used for joining two polymer composite parts and polymer composite to plastic parts; it is not suitable for joining composite to metals, alloys and ceramics.
*   Design of susceptor or implant is difficult for complicated shapes.

13.15   CONCLUSIONS

For the last few decades, composite materials are in great demand. They are used in structural and non-structural applications. Therefore, development of new products with intricate shapes requires joining of individual components in an assembly, and joining of composites is a challenging task. The following salient points can be concluded from the ongoing discussion:

*   Thermoplastics and thermosets behave differently during joining process. Thermosets are not suited for fusion welding. Only adhesive and mechanical joining are used for thermoset matrix composites. Fusion welding and

mechanical joining are used for thermoplastic matrix composites. In case of adhesive joining of thermoplastics, extra surface modification such as corona discharge treatment, plasma treatment or ion treatment is required.

- Surface energy, surface roughness and chemical properties of adherend surfaces play an important role in adhesive joining.
- Amount of heat required for fusion welding of thermoplastic matrix depends on the amorphous and crystalline phases present in thermoplastics.
- Chemical bonding, chain entanglement and chain diffusion are the basic mechanisms of bond formation in fusion welding.

REFERENCES

Awaja, F. 2016. Autohesion of polymers. *Polymer* 97: 387–407.

Bajpai, P. K., Singh, I., and Madaan, J. 2012. Joining of natural fiber reinforced composites using microwave energy: Experimental and finite element study. *Materials & Design* 35: 596–602.

Baldan, A. 2004. Review: Adhesively-bonded joints and repairs in metallic alloys, polymers and composite materials: Adhesives, adhesion theories and surface pretreatment. *Journal of Materials Science* 39: 1–49.

Davies, P., Cantwell, W. J., Jar, B., and Kaush, H. H. 1991. Joining and repair of a carbon fibre-reinforced thermoplastic. *Composites* 22: 425–431.

Fink, B. K., Mccullough, R. L., and Gillespie, J. 1995. A Model to predict the through thickness distribution of heat generation in cross-ply carbon fiber composites subjected to alternating magnetic field. *Centre for Composite Materials and Materials Science Program* 55: 119–130.

Kathirgamanathan, P. 1993. Microwave welding of thermoplastics using inherently conducting polymers. *Polymer* 34: 3405–3406.

Konig, W. C., Wulf, P. and Gral, H. W. 1985. Quality definition and assessment in drilling of fibre reinforced thermosets. *Annals of the CIRP* 38: 119–124.

Martinelli, M., Rolla, P. A., and Tombari, E. 1985. A method for dynamic dielectric measurements at microwave frequencies: Applications to polymerization process studies. *IEEE Transactions on Instrumentation and Measurement* IM-34(3): 417–421.

Messler, Jr., R. W. 2004. *Joining of Materials and Structures.* MA, USA: Elsevier Butterworth-Heinemann.

Mishra, R. and Sharma, A. K. 2016. Microwave material interaction phenomena: Heating mechanisms, challenges and opportunities in material processing Review. *Composites: Part A* 81: 78–97.

Rudolf, R., Mitschang, P., and Neitzel, M. 2000. Induction heating of continuous carbon-fibre-reinforced thermoplastics. *Composites: Part A* 31: 1191–1202.

Singh, I., Bajpai, P. K., Malik, D., Sharma, A. K., and Kumar, P. 2013. Feasibility study on microwave joining of green composites. *Akademeia* 1: 1–6.

Warren, K. C., Lopez-Anido, R. A., Freund, A. L., and Dagher, H. J. 2016. Resistance welding of glass fiber reinforced PET: Effect of weld pressure and heating element geometry. *Journal of Reinforced Plastics and Composites* 35: 974–985.

Yarlagadda, S. and Chai, T. C. 1998. An investigation into welding of engineering thermoplastics using focused microwave energy. *Journal of Materials Processing Technology* 74: 199–212.

Yarlagadda, S. and Kim, H. J. 2002. A study on the induction heating of conductive fiber reinforced composites. *Journal of Composite Materials* 36: 401–421.

Index

Note: Page numbers followed by f and t refer to figures and tables, respectively.

Printed and bound by CPI Group (UK) Ltd, Croydon, CR0 4YY

01/11/2024

01782622-0002